THE ANSWER BOOK

PRACTICAL MATHEMATICS

CONSUMER APPLICATIONS

Holt, Rinehart and Winston, Inc.

Austin   New York   San Diego   Chicago   Toronto   Montreal

Authors

Marguerite M. Fredrick
Computer Consultant, K–12
Pleasantville Union Free School District
Pleasantville, New York
Formerly Math Chairperson, 7–12
The Ursuline School
New Rochelle, New York

Steven J. Leinwand
Mathematics Consultant, K–12
Connecticut State Department of Education
Hartford, Connecticut

Robert D. Postman
Professor and Chairperson
Mathematics Education
Mercy College
Westchester, New York

Laurence R. Wantuck
Mathematics Supervisor, K–12
Broward County Schools
Fort Lauderdale, Florida

ISBN 0-03-012777-7

789-085-9876

CONTENTS

his book contains answers to the following blackline masters in the
eacher's Resource Package.

ractice Sheets   1–8

eteaching Sheets   8–12

pplications Sheets   12–13

roup Projects   13

ests
- Skills Inventory   13
- Cumulative Tests   14
- Chapter Posttests (Forms A and B)   14–16
  - Chapter 1   14
  - Chapter 2   14
  - Chapter 3   14
  - Chapter 4   14
  - Chapter 5   14
  - Chapter 6   14
  - Chapter 7   14
  - Chapter 8   14–15
  - Chapter 9   15
  - Chapter 10   15
  - Chapter 11   15
  - Chapter 12   15
  - Chapter 13   15
  - Chapter 14   15
  - Chapter 15   15
  - Chapter 16   15–16
  - Chapter 17   16
  - Chapter 18   16
- Final Test   16

PRACTICE SHEETS

PAGE 1 **1.** 187 **2.** 634 **3.** 781 **4.** 3,977 **5.** 6,825
6. 5,809 **7.** 41 **8.** 4.91 **9.** 10.33 **10.** 8.56 **11.** 1.14
12. 77.37 **13.** 454 **14.** 2,962 **15.** 5,545 **16.** 14,389
17. 9,609 **18.** 8,212 **19.** 15.4 **20.** 17.5 **21.** 112.4
22. 29.92 **23.** 31.7 **24.** 50.59 **25.** 10,864 **26.** 6,881
27. 7,270 **28.** 87.5 **29.** 9.9 **30.** 31.08 **31.** 7,812
32. 9,937 **33.** 6,010 **34.** 4,891 **35.** 6,208 **36.** 7,844
37. 52.5 **38.** 32 **39.** 45.92 **40.** 39.8 **41.** 34.9 **42.** 41.66
43. 46.74 **44.** 497 situps **45.** 2,059 mi **46.** 23.86 in.

PAGE 2 **1.** 404 **2.** 325 **3.** 758 **4.** 4,376 **5.** 912
6. 9,041 **7.** 857 **8.** 4,232 **9.** 4,164 **10.** 1,515 **11.** 981
12. 6,771 **13.** 1.29 **14.** 0.46 **15.** 6.9 **16.** 4.16 **17.** 78.42
18. 0.68 **19.** 1.93 **20.** 11.71 **21.** 0.23 **22.** 18.5
23. 27.03 **24.** 10.87 **25.** 211 **26.** 28 **27.** 364 **28.** 1,589
29. 7,782 **30.** 4,791 **31.** 3,879 **32.** 8,757 **33.** 1,975
34. 27.8 **35.** 48.71 **36.** 8.5 **37.** 18.16 **38.** 31.05
39. 28.54 **40.** 0.46 **41.** 81.5 **42.** 4.14 **43.** 1,070 yd
44. $418.00 **45.** $2.59 **46.** $135.29 **47.** 34.2 in.
48. 7 in.

PAGE 3 **1.** 112 **2.** 63 **3.** 600 **4.** 172 **5.** 324 **6.** 364
7. 396 **8.** 648 **9.** 1,228 **10.** 2,215 **11.** 4,242 **12.** 1,566
13. 17.2 **14.** 3.8 **15.** 2.1 **16.** 65.6 **17.** 4.2 **18.** 38
19. 7.34 **20.** 6.24 **21.** 2.87 **22.** 2.24 **23.** 40.15
24. 11.52 **25.** 114 **26.** 142 **27.** 192 **28.** 102 **29.** 886
30. 1,485 **31.** 872 **32.** 2,586 **33.** 20.7 **34.** 37 **35.** 49.8
36. 15.3 **37.** 8.46 **38.** 24.4 **39.** 1.66 **40.** 44.8 **41.** 540
42. 515 **43.** 4,856 **44.** 418 **45.** 7.21 **46.** 9.27 **47.** 18.16
48. 0.18 **49.** 35.45 **50.** 0.32 **51.** 81.09 **52.** 526 mi
53. $2.67 **54.** $15.75 **55.** 427.5 lb

PAGE 4 **1.** 16 **2.** 15 **3.** 29 R3 **4.** 73 **5.** 96 R1 **6.** 175
7. 31 **8.** 81 R2 **9.** 608 **10.** 510 **11.** 4,417 **12.** 927 R4
13. 871 R4 **14.** 569 R3 **15.** 408 R1 **16.** 6.1 **17.** 1.57
18. 1.18 **19.** 0.92 **20.** 0.76 **21.** 14 **22.** 262 **23.** 716 R2
24. 0.98 **25.** 0.75 **26.** 3.54 **27.** 4.3 **28.** 0.9 **29.** 1.3
30. 3.9 **31.** 0.6 **32.** 1.1 **33.** 3.2 **34.** 0.4 **35.** 0.6
36. 1.71 **37.** 0.78 **38.** 4.74 **39.** 17.23 **40.** 0.26 **41.** 2.82
42. 0.90 **43.** 0.25 **44.** 2.93 **45.** $8.72 **46.** 330 mi
47. $37.42 **48.** 76 mornings **49.** 2,500 copies **50.** 14
weeks

PAGE 5 **1.** 34% **2.** 15% **3.** 9% **4.** 275% **5.** 800%
6. 48% **7.** 3% **8.** 3.5% **9.** 2.6% **10.** 40% **11.** 700%
12. 804% **13.** 0.5; 50% **14.** 0.6; 60% **15.** 0.125; 12.5%
16. 0.75; 75% **17.** 0.3; 30% **18.** 0.625%; 62.5%
19. 2.875; 287.5% **20.** 6.25; 625% **21.** 3.8; 380%
22. 5.375; 537.5% **23.** 9.4; 940% **24.** 7.7; 770%
25. 0.37 **26.** 0.72 **27.** 0.02 **28.** 0.036 **29.** 0.005
30. 0.0009 **31.** 4 **32.** 5.09 **33.** $\frac{3}{10}$ **34.** $\frac{4}{5}$ **35.** $\frac{2}{5}$
36. $\frac{13}{20}$ **37.** $\frac{3}{20}$ **38.** $1\frac{1}{5}$ **39.** $1\frac{1}{2}$ **40.** $1\frac{1}{4}$ **41.** 0.25
42. $\frac{17}{20}$ **43.** 12% **44.** $\frac{3}{25}$ **45.** $1\frac{7}{25}$ **46.** 1.28 **47.** 18%
48. 0.18 **49.** 40% **50.** 0.4

PAGE 6 **1.** 5 **2.** 12 **3.** 160 **4.** 24 **5.** 10 **6.** 168 **7.** 37.8
8. 51.2 **9.** 30.5 **10.** 3.75 **11.** 3.4 **12.** 3.85 **13.** 7 **14.** 8
15. 18 **16.** 24 **17.** 27 **18.** 45 **19.** 44 **20.** 30 **21.** 14
22. 35 **23.** 25 **24.** 70 **25.** 22 **26.** 22.4 **27.** 18.5
28. 12.6 **29.** 20 **30.** 7.5 **31.** 3.5 **32.** 26 **33.** 27 **34.** 55

35. 250 **36.** 23 **37.** 48 patches **38.** 150 mi **39.** $70
40. $7.50 **41.** $48; $32 **42.** 99 students; 297 students

PAGE 7 **1.** 150 CD changers **2.** 275 VCR's **3.** 75 video
cameras **4.** 350 VCR's and video cameras **5.** Eddie
6. 4 VCR's **7.** 99 VCR's **8.** April **9.** 55 video cameras
10. 30 cameras **11.** $\frac{1}{5}$; 20% **12.** Check students'
graphs.

PAGE 8 **1.** 381 **2.** 37 **3.** 236 **4.** 16 **5.** 3.8 **6.** 6.4 **7.** 5.2
8. 8.8 **9.** 18 **10.** 10.5 **11.** 49 **12.** 72.5 **13.** 3.8 **14.** 3.7
15. 6.3 **16.** 6.4 **17.** 2 **18.** 12 **19.** 2.9; 5.2 **20.** no mode
21. 5.7 **22.** 8. **23.** 85; 92; none **24.** 5.8; 4.3; none
25. 54; 54; 54 **26.** 323.5; 315; 315 **27.** 14.2; 13.5; none
28. 72.4; 79; none **29.** 324.4; 286; 286 **30.** 4.8; 4.2; 4.2
31. 16.9; 17.2; none **32.** 43.4; 46; 35 **33.** 16.4; 18
34. 9; 8.8; 8.3 **35.** 15.5; 16; 17 **36.** 6; 6; none

PAGE 9 **1.** 1,296,854 **2.** 859,347 **3.** 20,862,366 **4.** 607
5. 5,303.625 **6.** 590.655 **7.** 324.19304 **8.** 10,356.624
9. 484.421 **10.** 63.5529 **11.** 203.5 **12.** 692 **13.** 67.2
14. 0.09 **15.** 18.8 **16.** 62.9 **17.** 119.6 **18.** 33.0 **19.** 16.3
20. 66.7 **21.** 39.6 **22.** 99.2 **23.** 239.3 **24.** 52.71
25. 0.15 **26.** 0.48 **27.** 146.24 **28.** 0.08 **29.** 1,720.59
30. 2.46 **31.** 0.57 **32.** 1.66 **33.** 147.253 **34.** 347.68
35. 749.935 **36.** 44.849 **37.** 6,032.88 **38.** 60.515
39. 5.6 **40.** 10.29 **41.** $17.36 **42.** $3.04 **43.** $12.50
44. $5.68

PAGE 10 **1.** 92 **2.** 83 **3.** 632 **4.** 708 **5.** 693 **6.** 83¢
7. $851 **8.** $7.14 **9.** 2.93 **10.** 8.31 **11.** 44 **12.** 48
13. 264 **14.** 433 **15.** 23¢ **16.** $265 **17.** $652 **18.** $2.17
19. $3.36 **20.** 3.56 **21.** 2.42 **22.** 3,240 **23.** 528,000
24. 457.8 **25.** 690 **26.** 4,800 **27.** 63,470 **28.** 510
29. 8.75 **30.** 52 **31.** 49 **32.** 0.84 **33.** 0.0056 **34.** 0.763
35. 0.869 **36.** 0.0075 **37.** 4.56 **38.** 68.50 **39.** 0.0009
40. 121 **41.** 8.85 **42.** 0.00087 **43.** $4.56 **44.** $2,995
45. $2.02 **46.** $1,450 **47.** 12.5 lb **48.** $27.69 **49.** $9.50
50. $25.03

PAGE 11 **1.** 70 **2.** 600 **3.** $11.30 **4.** 460 **5.** 17,000
6. 97,000 **7.** $9.00 **8.** $6.80 **9.** 11,000 **10.** 14 **11.** 640
12. 660 **13.** 400 **14.** 590 **15.** 640 **16.** 5,000 **17.** 2,200
18. 20,000 **19.** 85,000 **20.** $2.00 **21.** $0.40 **22.** 10,000
23. $39.10 **24.** $65.00 **25.** 64,000 **26.** $960 **27.** $0.08
28. 42,000 **29.** 0.39 **30.** 6.93 **31.** $591 **32.** 0.05
33. 9.8 **34.** 35.7 **35.** $50 **36.** $2.00 **37.** $200.00
38. $100 **39.** 120,000 **40.** 70,000 **41.** 10,000
42. 20,000

PAGE 12 **1.** 240 **2.** 900 **3.** 240,000 **4.** 140,000 **5.** $24
6. $1,400 **7.** $9,000 **8.** $6,000 **9.** 63 **10.** 450 **11.** 3,600
12. 40 **13.** 240 **14.** 350 **15.** $24 **16.** $350 **17.** 21
18. 200 **19.** 0.24 **20.** 0.06 **21.** 0.016 **22.** 20 **23.** 40
24. 8 **25.** 5 **26.** 200 **27.** 1.5 **28.** 20 **29.** 4 **30.** $400
31. $2 **32.** $30 **33.** $5 **34.** 4 **35.** 30 **36.** 200 **37.** 30
38. 2 **39.** 10 **40.** 300 **41.** 20 **42.** 10 **43.** $16.00
44. $6.00 **45.** 10 erasers **46.** 10 pens **47.** $15 **48.** $8
49. 3 sweatshirts **50.** 5 uniforms

PAGE 13 **1.** 16 **2.** 11 **3.** 24 **4.** 82 **5.** 14 **6.** 9 **7.** 11
8. 21 **9.** 15 **10.** 7 **11.** 4 **12.** 0 **13.** 16 **14.** 35 **15.** 75

16. 13 **17.** 58 **18.** 39 **19.** 15 **20.** 54 **21.** 36 **22.** 700
23. 600 **24.** 50 **25.** 300 **26.** 400 **27.** 1,800 **28.** 4 **29.** 7
30. 10 **31.** 25 **32.** 200 **33.** 500 **34.** 2,000 **35.** 8 **36.** 2
37. 2 mi **38.** 4 mi **39.** 9 mi **40.** 3 mi **41.** 55 h **42.** 12 in.

PAGE 14 **1.** b **2.** c **3.** b **4.** a **5.** c **6.** a **7.** b **8.** c **9.** c
10. a **11.** b **12.** a **13.** 300 mi **14.** 30 mi per gal
15. $20.00 **16.** 7 h per mo **17.** no **18.** yes **19.** no
20. yes

PAGE 15 Answers may vary. Possible answers are
given. **1.** calculator or paper and pencil **2.** mental
computation **3.** calculator or paper and pencil **4.** mental
computation or calculator **5.** calculator or paper and
pencil **6.** calculator or paper and pencil **7.** mental
computation **8.** calculator or paper and pencil **9.** $10.72;
paper and pencil or calculator **10.** $5; mental
computation **11.** $5.21; paper and pencil or calculator
12. $10; mental computation **13.** $16.50 **14.** $90.00
15. $5.16 **16.** $13.90

PAGE 16 **1.** $42 **2.** $52.50 **3.** $49.50 **4.** $88 **5.** $16.50
6. $40 **7.** $62.50 **8.** $66 **9.** $9.50 **10.** $57 **11.** $76
12. $54 **13.** $117.50 **14.** $196

PAGE 17 **1.** $4.15 **2.** $33.00 **3.** $140 **4.** Lakeside Park
Program **5.** Sallie's Snacks **6.** $3.33 **7.** 3 h **8.** $498
9. Best Burgers; Ford's Toys; Hilltop Day Camp
10. $81.25 **11.** $4.25 **12.** Hilltop Day Camp **13.** $61.50
14. Ford's Toys **15.** Camp counselor **16.** Answers may
vary.

PAGE 18 **1.** $7.16 **2.** $2.25 **3.** $0.95 **4.** $3.30 **5.** $0.70
6. $1.20 **7.** $12.60 **8.** $25.50 **9.** $38.10 **10.** $10.80
11. $24 **12.** $34.80 **13.** $14.40 **14.** $30 **15.** $44.40
16. $9.25 **17.** $36 **18.** $45.25 **19.** $11.10 **20.** $57
21. $68.10 **22.** $12.35 **23.** $63 **24.** $75.35 **25.** $9.50
26. $40.50 **27.** $50 **28.** $14 **29.** $79.50 **30.** $93.50
31. $12.60 **32.** $48 **33.** $60.60 **34.** $14.30 **35.** $72
36. $86.30

PAGE 19 **1.** $30.04 **2.** $60.08 **3.** $67.59 **4.** $37.55
5. $45.06 **6.** $21.03 **7.** $55.95 **8.** $48.82 **9.** $29.66
10. $64.96 **11.** $405.54 **12.** $240.32 **13.** $398.03
14. $210.28 **15.** $488.15 **16.** $4.27 **17.** $4.83
18. $13.92 **19.** $24.50 **20.** $34.93 **21.** $21.60
22. $127.50 **23.** $149.10 **24.** $11.20 **25.** $27 **26.** $120
27. $147 **28.** $11.04 **29.** $18 **30.** $97.50 **31.** $115.50
32. $8.67 **33.** $33.30 **34.** $135 **35.** $168.30 **36.** $12.64
37. $27.75 **38.** $105 **39.** $132.75 **40.** $9.97 **41.** $38
42. $180 **43.** $218 **44.** $16.37 **45.** $34.20 **46.** $157.50
47. $191.70 **48.** $14.40 **49.** $42 **50.** $202.50
51. $244.50 **52.** $18.36 **53.** $42 **54.** $189 **55.** $231
56. $17.35 **57.** $45.10 **58.** $216 **59.** $261.10
60. $19.61 **61.** $67.50 **62.** $384 **63.** $451.50
64. $33.91 **65.** $78 **66.** $513 **67.** $591 **68.** $44.38
69. $102.50 **70.** $600 **71.** $702.50 **72.** $52.76

PAGE 20 **1.** c **2.** b **3.** 75 h; 90 h **4.** 5 quarters, 5 dimes
5. $42; $24 **6.** 21 h, 63 h **7.** 8 quarters,10 dimes, 3
nickels **8.** $3\frac{3}{4}$ h; $4\frac{1}{2}$ h **9.** 7 quarters, 9 dimes, 5 pennies
10. 24 h, 48 h **11.** 25 h **12.** $12 per lesson

PAGE 21 **1.** W; F; 4:30–8:30 **2.** F 4–9; S 10–5 **3.** 8 h
4. 13 h **5.** $4.50 **6.** $4.05 **7.** $53.30 **8.** $48.60
9. Kitchen help **10.** Sales help **11.** Danny's Diner
12. Barbara's Bookstore **13.** Ray's Music Shop **14.** Ray's
Music Shop **15.** Barbara's Bookstore **16.** Danny's Diner
17. Answers may vary. **18.** Answers may vary.

PAGE 22 **1.** $284.80 **2.** $117.60 **3.** $112.85
4. $672.00 **5.** $144.55 **6.** $673.20 **7.** $254.93
8. $449.60 **9.** $118.02 **10.** $567.62 **11.** $152.32 **12.** $0
13. $152.32 **14.** $509.60 **15.** $224.55 **16.** $734.15
17. $308.80 **18.** $28.95 **19.** $337.75 **20.** $685.60
21. $347.09 **22.** $1,032.69 **23.** $587.20 **24.** $192.68
25. $779.88 **26.** $7.84 **27.** $11.60 **28.** $8.12 **29.** $5.92
30. 16 h **31.** 37 h

PAGE 23 **1.** 8:00 **2.** 4:00 **3.** 8 h **4.** 7:45 **5.** 3:30 **6.** $7\frac{3}{4}$
7. 8:45 **8.** 5:15 **9.** $8\frac{1}{2}$ h **10.** 6:45 **11.** 2:15 **12.** $7\frac{1}{2}$ h
13. 9:00 **14.** 5:15 **15.** $8\frac{1}{4}$ h **16.** Check students' time
sheets. Total: $31\frac{3}{4}$ h **17.** Check students' time sheets.
Total: $31\frac{1}{4}$ h **18.** $247.65 **19.** $197.50 **20.** $37\frac{1}{4}$ h;
$293.53 **21.** 37 h; $193.88

PAGE 24 **1.** $175.48 **2.** $245.38 **3.** $598.08
4. $859.17 **5.** $1,947.92 **6.** $19,994.00 **7.** $17,566.90
8. $10,532.16 **9.** $22,259.40 **10.** $11,810.64
11. Job A; $4,600 **12.** Job A; $4,656 **13.** Job B; $2,892
14. Job B; $1,741 **15.** Job B; $2,305 **16.** Job A; $5,890
17. $17,650 per y; $6,082 **18.** $4,060; $78.08
19. $14.50 **20.** $3,034

PAGE 25 **1.** $56.00 **2.** $21.75 **3.** $507.50 **4.** $720.00
5. $2,976.00 **6.** $83.20 **7.** $725.40 **8.** $530.10
9. $66.60 **10.** $90.85 **11.** 8 illustrations **12.** $480.00
13. $693.75 **14.** $168.30 **15.** 38 telegrams
16. 58 deliveries **17.** 13 rooms; $32\frac{1}{2}$ h **18.** 34 h; 408
ties

PAGE 26 **1.** $237.85 **2.** $175.44 **3.** $1,672.40
4. $1,733.40 **5.** $420.35 **6.** $55.86 **7.** $676.60
8. $154.86 **9.** $2,117.30 **10.** $9,159.90 **11.** $42.46
12. $238.50 **13.** $291.60 **14.** $8,400 **15.** $707.85
16. $400.50 **17.** $1,590.10

PAGE 27 **1.** $290.55 **2.** $209.77 **3.** $151.05
4. $111.98 **5.** $308.13 **6.** $219.31 **7.** $256.50
8. $183.81 **9.** $168.00 **10.** $56.33 **11.** $313.05
12. $381.80 **13.** $56.30 **14.** $21.92 **15.** $11.07
16. $6.53 **17.** $285.98 **18.** $97.92 **19.** $195.48
20. 80% **21.** 79% **22.** 83% **23.** $492.50

PAGE 28 **1.** $924 **2.** $496 **3.** $0 **4.** $85 **5.** $229.60
6. $132.40 **7.** $2,658.75 **8.** $986.25 **9.** $720.90
10. $145.10 **11.** $152.60 **12.** $140.40 **13.** $2,969.20
14. $1,698.80 **15.** $21.37 **16.** $256.44 **17.** $20.12
18. $241.44 **19.** $28.12 **20.** $337.44 **21.** $7.62
22. $91.44 **23.** $13.86 **24.** $166.32 **25.** $25.35
26. $304.20 **27.** $40.46 **28.** $551.85 **29.** $1,349.40
30. $488.23

PAGE 29 **1.** $209.40 **2.** $349 **3.** $693.75 **4.** $204.75
5. $1,260 **6.** $621.50 **7.** $310.50 **8.** $480.90 **9.** $691.60
10. $552.00 **11.** $368.55 **12.** $712.25 **13.** $247.50
14. $3,735

PAGE 30 **1.** Term **2.** Term **3.** Straight life **4.** $5.98
5. $3.40 **6.** $13.20 **7.** 20 y **8.** 10 y **9.** life **10.** no **11.** no
12. yes **13.** no **14.** yes **15.** not applicable **16.** $299;
$170; $660 **17.** Ansonia **18.** $13,875 **19.** $975
20. less expensive

PAGE 31 **1.** $20.75; $4.25 **2.** $22.75; $7.25 **3.** $22; $8 **4.** $26; $4 **5.** $29.75; $0.25 **6.** $22.50; $4.50 **7.** $93.75; $6.25 **8.** $89.95; $10.05 **9.** $13.50 **10.** $75

PAGE 32 **1.** $15.25; pay separately **2.** $20.50; buy all-day pass **3.** $23.50; buy all-day pass **4.** 7 times **5.** 8 times **6.** 12 times **7.** 3 times **8.** $47.25; $2.75 **9.** $56.40; $3.60 **10.** $68.60; $1.40 **11.** $6.40 **12.** $18 **13.** $5 **14.** $7.35 **15.** no **16.** no

PAGE 33 **1.** $27 **2.** $8.05 **3.** $47.50 **4.** $43.20 **5.** $24.80 **6.** $32.90 **7.** 5 tubes; $3.75 **8.** 8 packets; $2 **9.** $34.89 **10.** $1.35 **11.** $58.44; $1.56 **12.** $68; $7 **13.** $81.37; $3.63 **14.** $63.70; $1.30

PAGE 34 **1.** $54 **2.** $111.30 **3.** $84.50 **4.** $6.40 **5.** $140.65 **6.** $302.50 **7.** $388.19 **8.** $132.50 **9.** $190.95 **10.** $85 **11.** $164.75 **12.** $222 **13.** $492.95 **14.** $238.75 **15.** $7.50 **16.** $321.90 **17.** 3 wk **18.** 6 wk **19.** 2 wk **20.** 8 wk **21.** 4 wk **22.** 3 wk

PAGE 35 **1.** $3.76 **2.** $4.61 **3.** $2.10 **4.** $3.60 **5.** $2.20 **6.** $4.80 **7.** $2.95 **8.** $5.20 **9.** $7.29 **10.** $7.29 **11.** $10.42 **12.** $13.22 **13.** $2.11 **14.** $5.31 **15.** $1.33 **16.** $1.02 **17.** $1.80 **18.** $1.03 **19.** $1.39 **20.** $1.25 **21.** $2.50 **22.** $2.00

PAGE 36 **1.** a **2.** b **3.** c **4.** b **5.** 14 boxes **6.** 20 towels **7.** $\frac{4}{5}$ **8.** 8 boxes **9.** 5 shorts **10.** $\frac{1}{3}$ **11.** 3 classes **12.** 10 pants **13.** 19 boxes **14.** $\frac{1}{4}$ **15.** 3 cartons **16.** 11 students

PAGE 37 **1.** indoor **2.** both **3.** $4.90 **4.** $6.20 **5.** 1 h **6.** 40 min **7.** $100.00 **8.** $150 **9.** Ranks 2nd **10.** Ranks 1st **11.** ice-skating **12.** roller-skating **13.** ice-skating indoors **14.** swimming **15.** Answers may vary. **16.** Answers may vary.

PAGE 38 **1.** $258.00 **2.** $87.36 **3.** $173.61 **4.** $76.16 **5.** $258.97 **6.** $16.63 **7.** 20% **8.** 35% **9.** 15% **10.** 18% **11.** 11% **12.** 40% **13.** $325.50 **14.** $109.20 **15.** $47.20 **16.** $154.11 **17.** $86.65 **18.** $378.13 **19.** $1,347.46 **20.** $110.06; $88.05

PAGE 39 **1.** $17.00 **2.** $16.23 **3.** $3.77 **4.** $30.00 **5.** $29.23 **6.** $0.77 **7.** $40.00 **8.** $38.32 **9.** $1.68 **10.** $28.00 **11.** $27.00 **12.** $3.00 **13.** $49.00 **14.** $46.76 **15.** $3.24 **16.** $10.35 **17.** $24.47 **18.** $76.00 **19.** $49.98 **20.** $27.02 **21.** $114.00 **22.** $65.97 **23.** $48.03 **24.** $132.00 **25.** $75.96 **26.** $56.04

PAGE 40 **1.** $1.45 **2.** $1.80 **3.** $4.72 **4.** $7.11 **5.** $17.07 **6.** $0.72 **7.** $17.10 **8.** $50.16 **9.** $81.25 **10.** $482.38 **11.** $1.40 **12.** $1.41 **13.** $3.42 **14.** $3.42 **15.** $0.78 **16.** $0.79 **17.** $2.61 **18.** $2.62 **19.** $0.72 **20.** $12.67 **21.** $1.73 **22.** $51.03 **23.** $7.33 **24.** $190.58 **25.** $38.99 **26.** $781.59 **27.** $156.84 **28.** $2,117.34 **29.** $897.27 **30.** $11,773.26

PAGE 41 **1.** $6.00 **2.** $7.00 **3.** $4.50 **4.** $9.50 **5.** $10.00 **6.** $16.50 **7.** $212.85 **8.** $168.85 **9.** $156.25 **10.** $128.95 **11.** $154.15 **12.** $233.95 **13.** Pleated cords; B4756 6200; taupe; 29; 1; 28.50; 28.50; 1 lb 10 oz; Oxford shirts; B4256 1001; burgundy; M; 3; 24.00; 72.00; 2 lb 4 oz; sweaters; B4258 4858; green; L; 2; 39.99; 79.98;

2 lb 12 oz; 180.48; 6 lb 10 oz; 8.25; 1.00; 9.02; 198.75 **14.** $30.02 **15.** $21.94

PAGE 42 **1.** does **2.** does not **3.** does not **4.** does **5.** does not **6.** does **7.** 27¢ per oz; 24.6¢ per oz; 3 oz for 74¢ **8.** 9.9¢ per oz; 10.8¢ per oz; 9 oz for 89¢ **9.** 19.4¢ per oz; 16.2¢ per oz; 26 oz for $4.20 **10.** 9.9¢ per oz; 10.3¢ per oz; 15 oz for $1.49 **11.** 77.3¢ per lb; 69.8¢ per lb; 5 lb for $3.49 **12.** 96.6¢ per lb; 98.6¢ per lb; 6.2 lb for $5.99 **13.** $2.13 per lb; $2.01 per lb; 4.8 lb for $9.65 **14.** $1.65 per L; $1.45 per L; 3.5 L for $5.09 **15.** $7.90 **16.** $19.82

PAGE 43 **1.** less than $6.00 **2.** less than $6.00 **3.** more than $6.00 **4.** more than $6.00 **5.** less than $6.00 **6.** more than $6.00 **7.** $7.19; $12.81 **8.** $7.74; $12.26 **9.** $8.05; $11.95 **10.** $13.65; $6.35 **11.** $15.79; $4.21 **12.** $15.36; $4.64 **13.** $2.81; $5.62 **14.** $3.81

PAGE 44 **1.** $2.68 **2.** $6.00 **3.** included **4.** $1.68 **5.** $4.50 **6.** included **7.** $0.55 **8.** $1.25 **9.** included **10.** $0.58 **11.** $1.78 **12.** $3.50 **13.** $5.49 **14.** $13.53 **15.** $29.21 **16.** 45 min **17.** 8 min **18.** 30 min **19.** Not very **20.** Very **21.** Average **22.** Home **23.** Pasta-To-Go **24.** $23.72 **25.** $15.68 **26.** $1,254.24 **27.** $2,466.88

PAGE 45 **1–5.** Check students' work. **6–12.** Check student's work. **6.** $198.75 **7.** $152.50 **8.** $284.80 **9.** $384.28 **10.** $426.53 **11.** $235.82 **12.** $570.97 **13.** $146.54 **14.** $73.53

PAGE 46 **1.** $521.18 **2.** $511.08 **3.** 476 **4.** $15.60 **5.** $5.50 **6.** $505.58 **7.** $154.85 **8.** $247.30 **9.** $529.79 **10.** $471.82

PAGE 47 **1–4.** Check students' work. **1.** $199.65 **2.** $333.59 **3.** $248.94 **4.** $822.46 **5–8.** Check students' work. **9.** $174.74 **10.** $59.23 **11.** $672.68 **12.** $57.83 **13.** $840.87 **14.** $171.89 **15.** $96.89 **16.** $120.56 **17.** $166.06 **18.** $111.06 **19.** $113.54

PAGE 48 **1.** $63 **2.** $413 **3.** $120 **4.** $720 **5.** $437.50 **6.** $1,687.50 **7.** $972 **8.** $5,022 **9.** $3,144 **10.** $9,694 **11.** $687.50 **12.** $3,187.50 **13.** $712.50 **14.** $1,662.50 **15.** $2,121.80 **16.** $121.80 **17.** $4,436.83 **18.** $436.83 **19.** $535.93 **20.** $35.93 **21.** $6,932.92 **22.** $432.92 **23.** $3,812.93 **24.** $312.93 **25.** $7,354.38 **26.** $354.38 **27.** $6,843.00 **28.** $6,047.55 **29.** $19,424.40 **30.** $5,968.71 **31.** $14,859.00 **32.** $10,786.50

PAGE 49 **1.** None **2.** $1,500 **3.** $2,500 **4.** $7.00 **5.** $6.50 **6.** $6.50 **7.** 20¢ **8.** 20¢ **9.** 25¢ **10.** no **11.** no **12.** yes; $5\frac{1}{2}$% **13.** NOW **14.** Money market **15.** $0 **16.** $8.90 **17.** $8.10; $97.20 **18.** $169.20

PAGE 50 **1.** 364 829 704 3 **2.** 9/20/90 **3.** May 14, 1993 **4.** $18.29 **5.** March 25, 1990 **6.** 136 410 289 5 **7.** $210.60 **8.** $140.75 **9.** $48.82 **10.** $1.01 **11.** $119.68 **12.** $50.00

PAGE 51 **1.** $105.20; $1.32; $148.58 **2.** $96.80; $1.45; $211.65 **3.** $72.89; $1.02; $110.90 **4.** $133.20; $1.60; $134.80 **5.** $122.03; $2.14; $335.62 **6.** $371.88; $4.46; $891.14 **7.** $4.87 **8.** $9.99 **9.** $14.53 **10.** $7.97 **11.** $28.47 **12.** $24.63 **13.** $0.31 **14.** $311.14

15. $312.46 **16.** $3.91 **17.** $395.32 **18.** $1,143.89
19. $11.08 **20.** $1,652.80 **21.** $621.85 **22.** $7.16
23. $877.39 **24.** $399.61 **25.** $5.00 **26.** $583.68
27. $1,024.96 **28.** $10.19 **29.** $1,076.78

PAGE 52 **1.** $0.93 **2.** $1.62 **3.** $1.63 **4.** $1.40 **5.** $2.24
6. $1.91 **7.** $3.02 **8.** $3.07 **9.** $3.99 **10.** $1.73
11. $2.48; $4,762.48 **12.** $2.80; $6,015.53 **13.** $0.31;
$589.31 **14.** $0.38; $763.38 **15.** $15.58; $1,357.78
16. $16.26; $640.26 **17.** $14.16; $1,920.26 **18.** $11.80;
$341.05 **19.** $9.13; $708.13 **20.** $9.17; $309.17

PAGE 53 **1.** $491.80 **2.** $16.80 **3.** $657.18 **4.** $37.18
5. $835.02 **6.** $50.02 **7.** $2,166.84 **8.** $306.84
9. $11.78 **10.** $137.26 **11.** $38.79 **12.** $79.87
13. $14.90 **14.** $64.98 **15.** $62.94 **16.** $58.83
17. $106.29 **18.** $59.46 **19.** $82.20 **20.** $4,864
21. $171.60 **22.** $2,796 **23.** $73.06 **24.** $6,895.80
25. $152.61 **26.** $8,193.20 **27.** $223.76 **28.** $4,121.60
29. $21.03 **30.** $2,836.80

PAGE 54 **1.** $91 **2.** $15 **3.** $150 **4.** $31 **5.** $234
6. $28.50 **7.** $441 **8.** $38.63 **9.** $628.40 **10.** $42.37
11. $840.75 **12.** $50.60 **13.** $18.50 **14.** $16.90
15. $22.75 **16.** $21.35 **17.** $28.45 **18.** $38.65
19. $59.25 **20.** $44.22 **21.** $139 **22.** $78.70 **23.** $310.50;
$14.51 **24.** $220.50; $22.75 **25.** $1,830; $65.00
26. $427.50; $32.50 **27.** $368.70; $23.70 **28.** $116.85;
$20.90

PAGE 55 **1.** $309; $28 **2.** $8.28 **3.** $3.72 **4.** $1.66
5. $3,240; $3,888 **6.** $411.94 **7.** $6,475 **8.** $7,350

PAGE 56 **1.** 13.5% **2.** 15.75% **3.** $332.46 **4.** $86.58
5. 60 **6.** 24 **7.** $1,386.90 **8.** $619.92 **9.** Payments fixed
10. Payments fixed **11.** Credit card **12.** Home equity
loan **13.** $766.98 **14.** $207.90

PAGE 57 **1.** $11,192 **2.** $12,330 **3.** $13,775
4. $15,120 **5.** $15,483 **6.** $17,022.75 **7.** $26,497.25
8. $15,046.01 **9.** $12,852.33 **10.** $12,299.25
11. $9,138.43

PAGE 58 **1.** $8,800 **2.** $11,372.40 **3.** $2,572.40
4. $14,872.40 **5.** $11,205 **6.** $13,984.32 **7.** $2,779.32
8. $16,634.32 **9.** $8,913 **10.** $10,996.20 **11.** $2,083.20
12. $11,671.20 **13.** $5,340 **14.** $5,969.52 **15.** $629.52
16. $11,519.52 **17.** $7,420 **18.** $8,906.88 **19.** $1,486.88
20. $13,206.88 **21.** $9,180 **22.** $10,986.84 **23.** $1,806.84
24. $16,686.84 **25.** $8,990 **26.** $17,179.20 **27.** $8,189.20
28. $23,979.20 **29.** $9,420 **30.** $11,559.84 **31.** $2,139.84
32. $14,659.84 **33.** $8,710 **34.** $10,951.68 **35.** $2,241.68
36. $15,951.68 **37.** $5,065 **38.** $5,987.52 **39.** $922.52
40. $11,937.52 **41.** $2,076.36; $13,651.36 **42.** $868.00;
$11,513.00 **43.** $1,708.00; $16,388.00 **44.** $1,198.40;
$10,648.40 **45.** $2,020.36; $17,455.36

PAGE 59 **1.** 222 **2.** 20 **3.** 6,172 **4.** 20 **5.** 4,866 **6.** 25
7. 2,890 **8.** 15 **9.** 75 **10.** 15 **11.** 6,200 **12.** 20 **13.** $30
14. $40 **15.** $30 **16.** $40 **17.** $30 **18.** $30 **19.** $40
20. $50 **21.** $40 **22.** $50 **23.** $30 **24.** $40 **25.** $555.12
26. $339.61 **27.** $587.13 **28.** $232.20 **29.** $6,270;
$10,230 **30.** $4,556.20; $7,433.80 **31.** $13,110; $21,390
32. $10,165; $16,585

PAGE 60 **1.** $509.75 **2.** $491.13 **3.** $528.54
4. $515.10 **5.** 1.20 **6.** 1.55 **7.** 2.60 **8.** $226.90
9. $198.45 **10.** $58.20 **11.** $483.55 **12.** 3.80

13. $1,837.49 **14.** $258.45 **15.** $186.34 **16.** $58.20
17. $502.99 **18.** 1.85 **19.** $930.53 **20.** $284.00
21. $186.34 **22.** $58.20 **23.** $528.54 **24.** 2.35
25. $1,242.07 **26.** $273.85 **27.** $142.68 **28.** $58.20
29. $474.73 **30.** 3.00 **31.** $1,424.19 **32.** $258.45
33. $198.45 **34.** $58.20 **35.** $515.10 **36.** 1.20
37. $618.12

PAGE 61 **1.** $215.40 **2.** $77.84 **3.** $390.32 **4.** $230.24
5. $284.20 **6.** $44.68 **7.** $305.56 **8.** $1,099.00 **9.** $80
10. 135 **11.** $48.60 **12.** $128.60 **13.** $175 **14.** 96
15. $30.72 **16.** $205.72 **17.** $112 **18.** 498 **19.** $139.44
20. $251.44 **21.** $49 **22.** 38 **23.** $15.96 **24.** $64.96
25. $40 **26.** 500 **27.** $180.00 **28.** $220.00 **29.** about
$168 **30.** about $329 **31.** $438 **32.** $494.40

PAGE 62 **1.** $13,708.65 **2.** $14,352.20 **3.** $335.18
4. $303.17 **5.** $883.48 **6.** $1,191.58 **7.** $650.00
8. $728.33 **9.** $5,698.16 **10.** $5,836.65 **11.** $0.29
12. $0.28 **13.** Driver: Good; Passenger: Moderate
14. Driver: Moderate; Passenger: Moderate **15.** Car Z
16. Car Z **17.** Car Y **18.** $278.70 **19.** $32.01
20. $165.58

PAGE 63 **1.** 368 mi **2.** 779 mi **3.** 1,095 mi **4.** 1,022 mi
5. 1,762 mi **6.** 3,000 mi **7.** 1,100 mi **8.** 1,200 mi **9.** 1,600
mi **10.** 2,800 mi **11.** 5 mi **12.** 188 mi **13.** 507 mi **14.** 2
mi **15.** 26 h **16.** 60 h **17.** 14 h **18.** 28 h **19.** 60 h **20.** 12
h **21.** 380 mi **22.** Chicago

PAGE 64 **1.** 3.5 mi **2.** 2 mi **3.** 4.5 mi **4.** 10 mi **5.** 3 mi
6. 3.5 mi **7.** 3 mi **8.** 9.5 mi **9.** 71 **10.** 171 **11.** Granite
Park **12.** Granite Park **13.** 5 mi **14.** 2 mi **15.** 2 mi **16.** 4
mi **17.** 3 mi **18.** 6.0 mi **19.** 275; 266 **20.** 266 to 15 to
171; 275 to 171

PAGE 65 **1.** 5:50 P.M. **2.** 6:10 P.M. **3.** 11:05 P.M.
4. 12:45 A.M. **5.** #145 **6.** #204 **7.** 2:05 P.M. **8.** 6 h 15
min **9.** 4 h 10 min **10.** 1 h **11.** 12:14 P.M. **12.** 11:44
A.M. **13.** 1:31 P.M. **14.** 1:55 P.M. **15.** Train #173 at
11:49 A.M. **16.** #451 **17.** 4 h 27 min **18.** 5 h 5 min **19.** 6
min **20.** 17 min **21.** Train #214, Bus #106; 9 h 15 min

PAGE 66 **1.** Hemisphere Airlines Flight #247 **2.** Central
Airlines Flight #164 **3.** Hemisphere Airlines Flight #542
4. Hemisphere Airlines Flight #542 **5.** 3 h 30 min **6.** 2 h
7. 2 h 45 min **8.** 4 h 15 min **9.** 4 h 55 min **10.** 2 h 47
min **11.** 2 h 3 min **12.** 2 h 55 min **13.** 3 h 18 min **14.** 2 h
11 min **15.** 2 h 58 min **16.** $1,425 **17.** $1,785 **18.** $777
19. $1,875 **20.** $311 **21.** $466

PAGE 67 **1.** Blue Line 5 stops **2.** Blue Line 7 stops to
Metro Center, change to Red Line and go 1 more stop
3. Blue Line 2 stops to Metro Center, change to Red Line
and go 3 more stops **4.** Blue Line 14 stops **5.** Blue Line 8
stops to Metro Center, change to Red Line and go 2 more
stops **6.** Blue Line 8 stops **7.** $1.10 **8.** $0.80 **9.** $1.50
10. $0.80 **11.** $0.80 **12.** $1.55 **13.** 4 min **14.** 12 min
15. 34 min **16.** 12 min **17.** $11.00 **18.** 200 min; 3 h 20
min

PAGE 68 **1.** $6.30 **2.** $6.70 **3.** $13.90 **4.** $21.45
5. $18.75 **6.** $14.20 **7.** $19.10 **8.** $35.35 **9.** $27.30
10. $7.60 **11.** $0.90 **12.** $1.20 **13.** $0.60 **14.** $0.75
15. $1.20 **16.** $1.80 **17.** $2.10 **18.** $2.70 **19.** $1.50
20. $3.00 **21.** $0.60 **22.** $1.35 **23.** $2.40 **24.** $1.95
25. $1.65 **26.** $3.75 **27.** $8.00 **28.** $5.50 **29.** $8.85
30. $3.35 **31.** $2.51

PAGE 69 **1.** $9.56 **2.** $5.31 **3.** $6.09 **4.** $18.53

5. $9.59 **6.** $30.81 **7.** $34.89 **8.** $35.59 **9.** $44.51
10. $40.61 **11.** $29.75 **12.** $21.46 **13.** $245.00
14. $540 **15.** $540 **16.** $2,160 **17.** $600 **18.** $2,430
19. $1,800 **20.** $986.40, or about $1,000 **21.** $2,256, or
about $2,300 **22.** $962.13, or about $1,000

PAGE 70 **1.** 9 h **2.** 90 h **3.** 6 d **4.** $365 **5.** $88.54
6. Airport parking **7.** Meals and 4 nights in a motel
8. Little **9.** Good **10.** Plane **11.** Car or bus **12.** $20.44
13. $219.50 **14.** $77.54 **15.** $1,345.96; no **16.** Answers
may vary.

PAGE 71 **1.** $10,514.02 **2.** $468.32 **3.** $10,982.34
4. $789.58 **5.** $374.39 **6.** $149.16 **7.** $1,507 **8.** $18,721
9. $32,926 **10.** $11,331 **11.** $53,331 **12.** $30,298
13. $31,574 **14.** $24,132 **15.** $27,878 **16.** $82,127
17. $65,095.90 **18.** $30,083.56 **19.** $14,283.09
20. $29,344.09 **21.** $23,077.65 **22.** $27,398.81

PAGE 72 **1.** $17,580; $12,810 **2.** $26,755; $20,415
3. $34,850; $28,510 **4.** $13,420; $7,080 **5.** $47,530;
$38,070 **6.** $24,750; $19,090 **7.** $26,210; $3,814
8. $14,470; $2,099 **9.** $32,965; $6,145 **10.** $20,977;
$3,993 **11.** $21,566; $3,785 **12.** $38,700; $7,083 **13.** owe
$567.11 **14.** refund $65.56 **15.** refund $310.31 **16.** owe
$373.32 **17.** owe $295.99 **18.** refund $119.89

PAGE 73 **1.** yes; $1,900 **2.** $241.68 **3.** yes; $2,540
4. $400 **5.** 23,125.35 **6.** 109.08 **7.** 23,234.43
8. 2,540.00 **9.** 20,694.43 **10.** 1,900 **11.** 18,794.43
12. 3,558.00 **13.** 3,001.00 **14.** 557 **15.** 0

PAGE 74 **1.** $23 **2.** $407 **3.** $482 **4.** $1,778 **5.** $1,256
6. $0 **7.** $84; itemized **8.** $449; itemized **9.** $416;
standard **10.** $565; itemized **11.** $2,360; itemized
12. $5,684; itemized

PAGE 75 **1.** $144.80 **2.** $644.15 **3.** $464.75
4. $1,209.19 **5.** $252.30 **6.** $1,659.19 **7.** $2,124.52
8. $2,531.68 **9.** $818.54 **10.** $225.17 **11.** $403.43
12. $976.76 **13.** $1,321.06 **14.** $740.72 **15.** $2,051.54
16. $464.19 **17.** $791.92 **18.** $498.63 **19.** $75.22 (0)
20. $1,750 **21.** $15,122.63 **22.** $304.29 **23.** $25.50 (0)
24. $3,102.24 **25.** $244.74 (R) **26.** $4,900
27. $52,749.87 **28.** $1,621.25 **29.** $77.79 (R)
30. $892.79 **31.** $3.66 (0) **32.** $5,650 **33.** $18,008.24
34. $405.29 **35.** $10.14 (0)

PAGE 76 **1.** $5,550 **2.** $3,450 **3.** $2,100 **4.** 1
5. $2,100 **6.** $240 **7.** $175 **8.** $1,075 **9.** $1,300 **10.** $50
11. Federal: $1,912.50; state: $765; city: $148.75

PAGE 77 **1.** Single **2.** Married, separate return
3. Married, joint return **4.** $18,453 **5.** $35,030
6. $51,087 **7.** $343 **8.** $1,160 **9.** $1,718 **10.** $0
11. $1,250 **12.** $3,800 **13.** $18,796 **14.** $34,940
15. $49,005 **16.** $1,900 **17.** $1,900 **18.** $7,600
19. $1,618 **20.** $2,383 **21.** $5,409 **22.** $2,540
23. $1,880 **24.** $3,760 **25.** A: 1040EZ; B: 1040A; C:
1040A **26.** A:1040EZ; B: 1040A; C: 1040A **27.** A: 1040EZ;
B: 1040; C: 1040 **28.** A: standard deduction; B: itemized
deductions; C: itemized deductions **29.** A: $2,560; B:
$2,383; C: $5,409 **30.** A: $4,460; B: $4,283; C: $13,009
31. A: $14,356; B: $30,657; C: $35,996 **32.** A: $2,084; B:
$7,264; C: $6,313

PAGE 78 **1.** $264.60 **2.** $304.92 **3.** $346.36

4. $473.76 **5.** $635.04 **6.** $378 **7.** $436.80 **8.** $599.20
9. $982; $274.96 **10.** $1,582; $442.96 **11.** $1,245;
$348.60 **12.** $1,375; $385 **13.** $1,316.67; $386.67
14. $1,741.67; $487.67 **15.** $345 **16.** $598 **17.** $415
18. $385 **19.** $470 **20.** $763 **21.** $1,485 **22.** $657
23. $1,822.50

PAGE 79 **1.** $76,900; $109,400 **2.** $71,960; $98,410
3. $82,950; $121,150 **4.** $87,600; $129,960 **5.** $89,232;
$97,782 **6.** $35,904; $39,784 **7.** $116,840; $138,515
8. $138,060; $181,160 **9.** $4,565 **10.** $50,215
11. $5,085.85 **12.** $77,740.85 **13.** $10,527
14. $106,227 **15.** $9,779.40 **16.** $91,274.40
17. $7,914.90 **18.** $83,294.90 **19.** $9,637.50
20. $73,887.50 **21.** $11,994.50 **22.** $116,294.50
23. $91,562.40 **24.** $35,443 **25.** $49,114.06
26. $55,280.50

PAGE 80 **1.** $8,650; $1,030 **2.** $7,420; $1,153
3. $24,750; $2,026 **4.** $17,700; $1,395 **5.** $4,755; $885
6. $12,575; $1,175 **7.** $2,000 **8.** $3,000 **9.** $2,300
10. $2,700 **11.** $3,700 **12.** $1,300 **13.** $1,700
14. $3,300 **15.** $4,000 **16.** $4,300 **17.** $8,940; $976
18. $9,360; $1,314 **19.** $3,000 **20.** $4,000 **21.** about
$300 or $330 **22.** about $600 or $560

PAGE 81 **1.** $545.37 **2.** $859.95 **3.** $729.14
4. $679.50 **5.** $850.53 **6.** $1,041.39 **7.** $14,100
8. $79,900 **9.** $852.53 **10.** $11,040 **11.** $99,360
12. $959.82 **13.** $17,300 **14.** $69,200 **15.** $604.81
16. $12,225 **17.** $69,275 **18.** $715.61 **19.** $31,360
20. $125,440 **21.** $1,414.96 **22.** $11,250 **23.** $101,250
24. $926.44 **25.** $2,680 **26.** $2,335 **27.** $2,389
28. $4,527.25 **29.** $3,366.25 **30.** $2,893.75

PAGE 82 **1.** $58,520 **2.** $77,690 **3.** $43,819
4. $80,400 **5.** $81,312 **6.** $2,242.56 **7.** $2,017.17
8. $2,036.34 **9.** $2,947.68 **10.** $2,169.96 **11.** $634
12. $750 **13.** $902 **14.** $926 **15.** $901 **16.** $77,120
17. $72,800 **18.** $96,000; $2,774.40 **19.** $846.20
20. $53,220; $2,192.66 **21.** $587.72 **22.** $59,400;
$1,882.98 **23.** $724.92

PAGE 83 **1.** $6,500 **2.** $32,500 **3.** $13,000 **4.** $3,250
5. $8,200 **6.** $41,000 **7.** $16,400 **8.** $4,100 **9.** $11,050
10. $55,250 **11.** $22,100 **12.** $5,525 **13.** $12,040
14. $60,200 **15.** $24,080 **16.** $6,020 **17.** $7,856
18. $39,280 **19.** $15,712 **20.** $3,928 **21.** $780
22. $1,560 **23.** $940 **24.** $1,880 **25.** $1,645 **26.** $3,290
27. $2,035 **28.** $4,070 **29.** $2,815.50 **30.** $5,631

PAGE 84 **1.** 34,712 kWh **2.** 18,905 kWh **3.** 14,722
kWh **4.** 60,253 kWh **5.** 1,465 **6.** $56.70 **7.** 1,076
8. $57.57 **9.** 12,937 **10.** $322.13 **11.** $72.38
12. $116.01 **13.** $87.22 **14.** $182.61 **15.** 3.216
16. $41.00 **17.** 2.491 **18.** $35.67 **19.** 4.709 **20.** $65.03
21. 5.693 **22.** $66.27 **23.** $0.006734 **24.** $0.01405
25. $367.14 **26.** $68.17

PAGE 85 **1.** 40 ft **2.** 58 ft **3.** 54 ft **4.** 72 ft **5.** 2 gal;
$25.90 **6.** 2 gal; $22.50 **7.** 5 gal; $53.75 **8.** $1,680
9. $4,230 **10.** $3,780 **11.** l = 12 in.; w = 10 in. **12.** l = 8
in.; w = 6 in. **13.** l = 12.5 cm; w = 9 cm **14.** 2 gal; $30.90
15. 4 gal; $58.20 **16.** $6,468 **17.** l = 60 cm; w = 55 cm
18. 35 square yd; $682.50 **19.** 30 square yd; $600

PAGE 86 **1.** 10.5% **2.** 11.5% **3.** Fixed **4.** Adjustable
5. $686.16 **6.** $731.52 **7.** $2,190 **8.** $1,965 **9.** 30 y

10. 20 y **11.** $172,627.20 (may vary) **12.** $219,456
13. People's Trust **14.** People's Trust **15.** Nationwide
Bank **16.** Downtown Bank **17.** $27,561.60 **18.** $262,128
19. $293,335.20

PAGE 87 **1.** $5,940 **2.** $6,600 **3.** $8,250 **4.** $9,900
5. $14,850 **6.** $16,500 **7.** $7,560 **8.** $8,400 **9.** $10,500
10. $12,600 **11.** $18,900 **12.** $21,000 **13.** 25 **14.** $8.75
15. 75 **16.** $21.75 **17.** 64 **18.** $18.24 **19.** 240
20. $74.40 **21.** 225 **22.** $74.25 **23.** $32.00 **24.** $57.60
25. $26.00 **26.** $135.60 **27.** $74.75 **28.** $129.00
29. $6.40 **30.** $461.35

PAGE 88 **1.** $1\frac{1}{4}$ in. **2.** $2\frac{5}{16}$ in. **3.** $3\frac{1}{4}$ in. **4.** 3 in. **5.** $1\frac{13}{16}$ in.
6. $1\frac{1}{4}$ in. **7.** $\frac{3}{16}$ in. **8.** $\frac{1}{16}$ in. **9.** 91 ft **10.** 5 **11.** 1
12. 266 ft 8 in. **13.** 14 **14.** 3 **15.** 156 ft **16.** 8 **17.** 2 ft
18. 183 ft **19.** 10 **20.** 2 **21.** 335 ft **22.** 17 **23.** 4
24. $1,065.90 **25.** $229.60 **26.** $265.80 **27.** $1,089.20

PAGE 89 **1.** 3 **2.** $94.50 **3.** 5 **4.** $214.75 **5.** 4
6. $131.80 **7.** 8 **8.** $324.40 **9.** 12 **10.** $465.00 **11.** 11
12. $480.15 **13.** 27 **14.** $1,059.75 **15.** $361.00
16. $833.25 **17.** $1,222.50 **18.** $1,023.41 **19.** $3,334.40
20. $3,710.00 **21.** $2,341.50 **22.** $3,492.00 **23.** $463.29
24. $516.49

PAGE 90 **1.** 96 **2.** 51 **3.** 45 **4.** 1 **5.** $17.50 **6.** 135
7. 51 **8.** 84 **9.** 1 **10.** $17.50 **11.** 360 **12.** 102 **13.** 258
14. 2 **15.** $35.00 **16.** 600 **17.** 102 **18.** 498 **19.** 3
20. $52.50 **21.** 1,824 **22.** 102 **23.** 1,722 **24.** 9
25. $157.50 **26.** $290.00 **27.** $252.00 **28.** $2,190.00
29. $3,150.00 **30.** $3,344.00 **31.** 120 square ft
32. 96 square ft **33.** 48 square ft **34.** 432 square ft
35. 2 gal

PAGE 91 **1.** 150 **2.** $90 **3.** 360 **4.** $216 **5.** 1,440
6. $864 **7.** 2,250 **8.** $1,350 **9.** 23,040 **10.** $13,824
11. 1,350 **12.** $202.50 **13.** 3 **14.** $22.50 **15.** 5,265
16. $789.75 **17.** 12 **18.** $90 **19.** 16,875 **20.** $2,531.25
21. 38 **22.** $285 **23.** 56,250 **24.** $8,437.50 **25.** 125
26. $937.50 **27.** 115,200 **28.** $17,280 **29.** 256
30. $1,920 **31.** 142,875 **32.** $21,431.25 **33.** 318
34. $2,385 **35.** 564 square ft; 372 square ft **36.** 8,424
bricks **37.** 19 bags

PAGE 92 **1.** 9 trees **2.** 16 ft **3.** 12 posts **4.** 47 ft
5. 208 ft wide by 110 ft long **6.** 28 trees

PAGE 93 **1.** $57.48 **2.** $69.95 **3.** Average **4.** Excellent
5. Washable **6.** Washable **7.** Money back **8.** 10 y
9. Regular **10.** Deluxe **11.** Good **12.** Deluxe **13.** $12.47;
The Good paint might not last 10 y and they might need to
repaint for $57.48 or more **14.** Answers may vary.

PAGE 94 **1.** 20.25 **2.** $397.31 **3.** 66.25 **4.** $1,299.83
5. 62 **6.** $1,216.44 **7.** 3.75 **8.** $73.58 **9.** 19 **10.** $372.78
11. 14 **12.** $382.76 **13.** 27.75 **14.** $758.69 **15.** 5.5
16. $150.37 **17.** 35 **18.** $956.90 **19.** 37.5 **20.** $1,025.25
21. 9 d **22.** 6 d **23.** 4 d **24.** 64 column-in.; $867.84
25. 55 column-in.; $875.60

PAGE 95 **1–9.** Answers may vary.

PAGE 96 **1.** 304 **2.** 7,185 **3.** 31 **4.** 91 **5.** 67 **6.** 7,595
7. 179,625 **8.** 1,680 **9.** 2,271 **10.** 775 **11.** 56,762
12. 224 **13.** 163 **14.** 59 **15.** 26,942 **16.** 614 **17.** 1,033
18. $38.97 **19.** $37.85 **20.** $140.32 **21.** $62.50

22. $253.16 **23.** $4.61 **24.** $29.03 **25.** $401.76
26. $37.42 **27.** $129.44

PAGE 97 **1.** 37 **2.** 85 **3.** 48 **4.** 114 **5.** 784 **6.** 211
7. 21 **8.** 180 **9.** 62 **10.** 18 **11.** 23 **12.** 6 **13.** 60 **14.** 608
15. 120 **16.** 118 **17.** 98 **18.** 27 albums **19.** 2 sweaters

PAGE 98 **1.** 4 **2.** 3.3 **3.** 0.8 **4.** 8.9 **5.** 3.13 **6.** 1.39
7. 1.38 **8.** 8 **9.** 6 **10.** 5.5 **11.** 4 **12.** 1.25 **13.** 2.5 **14.** 65
drops per min **15.** 275 drops per min **16.** 25 drops per
min **17.** 120 mL **18.** 160 mL **19.** 144 mL

PAGE 99 **1.** 23 min 26 s **2.** 18 min 28 s **3.** 26 min 27 s
4. 15 min 44 s **5.** 20 min 1 s **6.** 25 min 21 s **7.** 6 min 34
s **8.** 11 min 32 s **9.** 3 min 33 s **10.** 14 min 16 s **11.** 9 min
59 s **12.** 4 min 39 s **13.** $145.80 **14.** $180.60 **15.** $750
16. $1,503.90 **17.** 30 spots **18.** 39 spots **19.** 31 spots
20. $177.30 **21.** $2,323.30 **22.** $486.80 **23.** $2,545.80
24. $382.50

PAGE 100 **1.** $340 **2.** $8,450 **3.** $9,724 **4.** $325
5. $411.40 **6.** $10,696.40 **7.** $11,700 **8.** $452.54
9. $11,766.04 **10.** $450 **11.** $11,700 **12.** $497.79
13. $575 **14.** $547.57 **15.** $14,236.82 **16.** $14,950
17. $602.33 **18.** $15,660.58 **19.** $700 **20.** $662.56
21. $17,226.56 **22.** $700 **23.** $18,200 **24.** $101,092.94
25. $106,600 **26.** Job 1 **27.** Job 1 **28.** Job 2 **29.** Job 2
30. Answers may vary. **31.** Answers may vary. Students
may decide that Rochelle should take Job 2. **32.** 40 h
33. 43 h **34.** 40 h **35.** $7.75 **36.** $8.89 **37.** $11.63
38. $8.89 **39.** Answers may vary.

PAGE 101 **1.** $1.00 **2.** 16.7% **3.** $2.50 **4.** 20.8%
5. $10.10 **6.** 11.8% **7.** $6.00 **8.** 8.3% **9.** $16.00
10. 10.1% **11.** $27.00 **12.** 8.5% **13.** $0.17 **14.** 17.3%
15. $1.50 **16.** 42.9% **17.** $1,800 **18.** $16,200
19. $1,470 **20.** $19,530 **21.** $1,880 **22.** $21,620
23. $3,084 **24.** $22,616 **25.** $2,416.50 **26.** $24,433.50
27. $2,698.75 **28.** $29,051.25 **29.** $26,815.80
30. 9.3% **31.** 19.4% **32.** $25,668.75 **33.** 15.0%
34. $7,879.62

PAGE 102 **1.** $198 **2.** $121 **3.** $92 **4.** $51 **5.** $72
6. $48 **7.** $13 **8.** $50 **9.** $70 **10.** $40 **11.** $52 **12.** $92
13. $85 **14.** $25 **15.** $28 **16.** $27 **17.** $18 **18.** $16
19. $22 **20.** $228 **21.** $86 **22.** $20 **23.** $20 **24.** $26
25. $65

PAGE 103 **1.** $92 **2.** $426 **3.** $215 **4.** $885 **5.** $496
6. $2,100 **7.** $175 **8.** $1,556 **9.** $184 **10.** $186
11. $173 **12.** $241; $118 **13.** $931; $241; $2,057; $317
14. $300; $516

PAGE 104 **1.** $276,830 **2.** $294,500 **3.** $321,860
4. $397,130 **5.** $369,400 **6.** $312,900 **7.** $450,740
8. $335,020 **9.** $331,390 **10.** $348,410 **11.** $386,220
12. $406,880 **13.** $465,050 **14.** $371,070 **15.** $84,160
16. $88,990 **17.** $20,660 **18.** $25,650

PAGE 105 **1.** $2,250 **2.** $1,356 **3.** $824 **4.** $8,540
5. $88,580 **6.** $104,430 **7.** $46,363 **8.** $58,067
9. $114,800 **10.** $18,980 **11.** $80,318 **12.** $145,670
13. $84,363 **14.** $61,307 **15.** $149,730 **16.** $90,457
17. $59,273

PAGE 106 **1.** $869 **2.** $1,119 **3.** $671 **4.** $671
5. $2,264 **6.** $1,640 **7.** $1,640 **8.** $1,990 **9.** 0 **10.** 0
11. 52 h **12.** $42\frac{1}{2}$ h **13.** Eliminate savings; cut in living

  ANSWER BOOK Practical Mathematics: Consumer Applications (Practice) expenses **14.** Away from home $7\frac{1}{2}$ h more per wk; cut in living expenses **15.** Plan 2 **16.** Answers may vary.

PAGE 107 **1.** $100 **2.** $200 **3.** $112.50 **4.** $750 **5.** $500 **6.** $7,500 **7.** $58.32 **8.** $583.20 **9.** $31.54 **10.** $144.56 **11.** $3,889 **12.** $9,350 **13.** $197.52 **14.** $47.52 **15.** $450.24 **16.** $50.24 **17.** $688.80 **18.** $188.80 **19.** $271.05 **20.** $83.55 **21.** $4,840.80 **22.** $1,840.80 **23.** $18,024 **24.** $8,024 **25.** $25,160 **26.** $12,660 **27.** $311.12 **28.** $111.12 **29.** $868.60 **30.** $368.60 **31.** $379.47 **32.** $116.97 **33.** $2,805 **34.** $1,305 **35.** $9,057.60 **36.** $4,557.60 **37.** $66,976 **38.** $26,976

PAGE 108 **1.** 8.52% **2.** 7% **3.** 8.10% **4.** 8.96% **5.** 8.80% **6.** $149.10 **7.** $176 **8.** $172.13 **9.** $35 **10.** $88 **11.** $352 **12.** $32.08 **13.** $129.60 **14.** $319.50 **15.** $182.25 **16.** $1,022.40 **17.** $93.33 **18.** $1,056; more; $33.60 **19.** $162; more; $68.67 **20.** $405 **21.** $426; more; $21

PAGE 109 **1.** $49\frac{1}{2}$ **2.** $23\frac{1}{4}$ **3.** 13,600 **4.** $3\frac{7}{8}$ **5.** $5.25 **6.** $23\frac{5}{8}$ **7.** $\frac{1}{4}$ **8.** $6,402.50 **9.** $0.80 **10.** $8.00 **11.** 2.7% **12.** 0 **13.** 0 **14.** 0 **15.** $2.20 **16.** $187 **17.** 9.4% **18.** $5.25 **19.** $2,100 **20.** 10.7% **21.** (P) $190 **22.** (L) $202.50 **23.** (L) $525

PAGE 110 **1.** Burton City **2.** Dale Township **3.** Dale Township **4.** Eagle Products **5.** Bachman Products; Burton City **6.** Crestway Metals; Dale Township; Eagle Products **7.** $6,453 **8.** $7,160 **9.** $24,600 **10.** $14,800 **11.** $26,262.50 **12.** $32,265 **13.** $28,640 **14.** $35,875 **15.** 97% **16.** 88.5% **17.** 82.5% **18.** 95.75% **19.** 98.9% **20.** 99.8% **21.** 118% **22.** 112.5% **23.** 120% **24.** 122.5% **25.** 125% **26.** 130% **27.** $540 **28.** $850 **29.** $1,250 **30.** $5,625 **31.** $1,143.75 **32.** $1,562.50 **33.** $3,400 **34.** $3,375 **35.** $2,812.50 **36.** $1,875 **37.** 8.5% **38.** 8.3% **39.** 5.8% **40.** 8.9%

PAGE 111 **1.** $1,846 **2.** $22,752 **3.** $7,852.50 **4.** $663.50 **5.** $24,264 **6.** $18,326 **7.** $4,983 **8.** $13.68 **9.** $273.60 **10.** $2,052 **11.** $16.09 **12.** $321.80 **13.** $2,413.50 **14.** $16.61 **15.** $332.20 **16.** $2,491.50 **17.** $30.34 **18.** $606.80 **19.** $4,551 **20.** $8.05 **21.** $161 **22.** $1,207.50 **23.** 199 shares **24.** 1,085 shares **25.** $900 **26.** $1,117.50 **27.** $552 **28.** $1,610

PAGE 112 **1.** 2.00 **2.** $11,000 **3.** $916.67 **4.** 1.67 **5.** $5,921.82 **6.** $493.49 **7.** 1.98 **8.** $7,318.08 **9.** $609.84 **10.** 1.57 **11.** $9,248.09 **12.** $770.67 **13.** 2.00 **14.** $15,260 **15.** $1,271.67 **16.** 1.70 **17.** $5,950 **18.** $495.83 **19.** 2.00 **20.** $12,992 **21.** $1,082.67 **22.** 1.87 **23.** $8,860.06 **24.** $783.34 **25.** 1.94 **26.** $7,618.38 **27.** $634.87 **28.** 1.76 **29.** $5,852 **30.** $487.67 **31.** 1.71 **32.** $5,868.72 **33.** $489.06 **34.** 1.89 **35.** $13,301.82 **36.** $1,108.49 **37.** 1.85 **38.** $13,247.85 **39.** $1,103.99 **40.** 1.98 **41.** $8,506.08 **42.** $708.84

PAGE 113 **1.** $18,000 **2.** $9,000 **3.** $22,500 **4.** $11,250 **5.** $18,000 **6.** $6,000 **7.** $12,500 **8.** $2,500 **9.** $27,000 **10.** $17,500 **11.** $4,000 **12.** $5,000

PAGE 114 **1.** $22.50 **2.** 4.4% **3.** (I); 8.9% **4.** High **5.** $75.67 **6.** $2.25 **7.** 3.0% **8.** (D); 24.4% **9.** None **10.** 0% **11.** (I); 10.2% **12.** Low **13.** $124.25 **14.** $4.50 **15.** (D); 2.4% **16.** High **17.** $16.50 **18.** $2.00 **19.** 12% **20.** Low **21.** $39.45 **22.** $0.60 **23.** 1.5% **24.** (I); 10.1% **25.** Moderate **26.** NeoElec **27.** Franklin **28.** State Gas **29.** Ray G & E **30.** Tor G & E **31.** State Gas **32.** NeoElec; T or G & E; low risk **33.** Answers may vary. An example: 536 shares of NeoElec and 38 shares of State Gas

PAGE 115 **1.** $\frac{3}{8}$ **2.** $\frac{1}{4}$ **3.** $\frac{1}{4}$ **4.** $\frac{1}{8}$ **5.** $\frac{5}{8}$ **6.** $\frac{5}{8}$ **7.** $\frac{3}{4}$ **8.** $\frac{1}{2}$ **9.** $\frac{3}{4}$ **10.** $\frac{1}{15}$ **11.** $\frac{4}{15}$ **12.** $\frac{1}{5}$ **13.** $\frac{2}{3}$ **14.** $\frac{2}{3}$ **15.** $\frac{7}{15}$ **16.** $\frac{8}{15}$ **17.** $\frac{1}{5}$ **18.** 0 **19.** $\frac{2}{5}$ **20.** 0 **21.** $\frac{14}{15}$ **22.** $\frac{1}{5}$ **23.** $\frac{11}{15}$ **24.** $\frac{1}{13}$ **25.** $\frac{1}{52}$ **26.** $\frac{2}{13}$ **27.** $\frac{5}{52}$ **28.** $\frac{12}{13}$ **29.** $\frac{1}{26}$ **30.** $\frac{2}{13}$ **31.** $\frac{1}{52}$ **32.** $\frac{1}{4}$ **33.** $\frac{3}{4}$ **34.** $\frac{1}{26}$ **35.** $\frac{1}{26}$ **36.** $\frac{12}{13}$ **37.** 0 **38.** $\frac{1}{2}$ **39.** $\frac{1}{6}$ **40.** $\frac{2}{3}$ **41.** $\frac{2}{3}$ **42.** $\frac{1}{2}$ **43.** $\frac{5}{6}$

PAGE 116 Check students' tree diagrams. **1.** (2,H); (2,T); (3,H); (3,T); (4,H); (4,T); (5,H); (5,T); (6,H); (6,T) **2.** (1,1); (1,2); (1,3); (1,4); (2,1); (2,2); (2,3); (2,4); (3,1); (3,2); (3,3); (3,4); (4,1); (4,2); (4,3); (4,4) **3.** (H,H,R); (H,H,B); (H,T,R); (H,T,B); (T,H,R); (T,H,B); (T,T,R); (T,T,B) **4.** (H,R); (H,B); (H,Y); (T,R); (T,B); (T,Y) **5.** (1,A); (1,B); (1,C); (2,A); (2,B); (2,C); (3,A); (3,B); (3,C) **6.** (1,2); (1,4); (1,6); (1,8); (3,2); (3,4); (3,6); (3,8); (5,2); (5,4); (5,6); (5,8) **7.** (A,A,1); (A,A,2); (A,A,3); (A,A,4); (A,B,1); (A,B,2); (A,B,3); (A,B,4); (B,A,1); (B,A,2); (B,A,3); (B,A,4); (B,B,1); (B,B,2); (B,B,3); (B,B,4) **8.** (H,H,H,1); (H,H,H,2); (H,H,T,1); (H,H,T,2); (H,T,H,1); (H,T,H,2); (H,T,T,1); (H,T,T,2); (T,H,H,1); (T,H,H,2); (T,H,T,1); (T,H,T,2); (T,T,H,1); (T,T,H,2); (T,T,T,1); (T,T,T,2) **9.** 60 different meals **10.** 6 different outfits **11.** 20 different ways **12.** 48 different lunches **13.** 24 combinations **14.** 12 different routes

PAGE 117 **1.** $\frac{1}{144}$ **2.** $\frac{11}{72}$ **3.** $\frac{1}{4}$ **4.** $\frac{1}{8}$ **5.** $\frac{1}{16}$ **6.** $\frac{3}{8}$ **7.** $\frac{1}{72}$ **8.** $\frac{5}{72}$ **9.** $\frac{1}{36}$ **10.** $\frac{1}{9}$ **11.** $\frac{5}{12}$ **12.** 0 **13.** $\frac{5}{36}$ **14.** $\frac{5}{12}$ **15.** $\frac{1}{18}$ **16.** $\frac{5}{9}$ **17.** $\frac{1}{9}$ **18.** $\frac{1}{6}$ **19.** $\frac{3}{25}$ **20.** $\frac{1}{50}$ **21.** $\frac{6}{25}$ **22.** $\frac{1}{4}$ **23.** $\frac{21}{50}$ **24.** $\frac{27}{100}$ **25.** $\frac{6}{25}$ **26.** $\frac{1}{25}$ **27.** $\frac{1}{2}$ **28.** $\frac{6}{25}$ **29.** $\frac{21}{100}$ **30.** 0

PAGE 118 **1.** independent **2.** independent **3.** dependent **4.** independent **5.** $\frac{3}{14}$ **6.** $\frac{5}{14}$ **7.** $\frac{1}{14}$ **8.** $\frac{2}{7}$ **9.** $\frac{3}{14}$ **10.** $\frac{3}{14}$ **11.** $\frac{1}{8}$ **12.** $\frac{15}{56}$ **13.** $\frac{1}{36}$ **14.** $\frac{1}{4}$ **15.** $\frac{1}{8}$ **16.** $\frac{1}{4}$ **17.** $\frac{1}{6}$ **18.** $\frac{7}{36}$ **19.** $\frac{3}{10}$ **20.** $\frac{1}{5}$ **21.** $\frac{1}{30}$ **22.** $\frac{2}{15}$ **23.** $\frac{1}{6}$ **24.** $\frac{1}{15}$

PAGE 119 **1.** 12 d **2.** 1 d **3.** 8 d **4.** 10 d **5.** 20 d **6.** 12 d **7.** 10 times **8.** 15 times **9.** 30 times **10.** 5 times **11.** 45 times **12.** 16.7% **13.** 8.3% **14.** 75% **15.** 15 games **16.** 23 games **17.** 38 games **18.** 27 games **19.** 45 games **20.** 87 games **21.** 30 games **22.** 68 games **23.** 180 winners

PAGE 120 **1.** JA, JS, JC, AS, AC, SC **2.** snowflakes, snowmen; snowflakes, snowballs; snowflakes, reindeer; snowmen, snowballs; snowmen, reindeer; snowballs, reindeer **3.** MHG, MHS, HGS; 4 combinations **4.** JLSV, JLVS, JSLV, JSVL, JVLS, JVSL; LJSV, LJVS, LSJV, LSVJ, LVJS, LVSJ; SJLV, SJVL, SLJV, SLVJ, SVJL, SVLJ; VJLS, VJSL, VLJS, VLSJ, VSJL, VSLJ; 24 arrangements **5.** 569, 596, 659, 695, 956, 965; 6 arrangements **6.** RR, BL, NN; RR, NN, BL; BL, RR, NN; BL, NN, RR; NN, RR, BL; NN, BL, RR; 6 arrangements **7.** 0, 1, 2; 0, 2, 1; 0, 3, 0; 1, 1, 1; 1, 2, 0; 2, 1, 0; 6 combinations **8.** 9:00 A.M.–3:00 P.M.,

9:00 A.M.–4:00 P.M., 9:00 A.M.–5:00 P.M., 10:00
A.M.–3:00 P.M., 10:00 A.M.–4:00 P.M., 10:00 A.M.–5:00
P.M., 11:00 A.M.–3:00 P.M., 11:00 A.M.–4:00 P.M.,
11:00 A.M.–5:00 P.M.; 9 combinations

PAGE 121 **1.** 10 **2.** 35 **3.** 6 **4.** 6 **5.** 5 **6.** 12 **7.** 13
8. 32 **9.** 50 **10.** 72 **11.** 3 **12.** 150 **13.** 26 **14.** 16 **15.** 11
16. 48 **17.** 10 **18.** 27 **19.** 47 **20.** 36 **21.** 52 **22.** 64
23. 8 **24.** 53 **25.** 26 **26.** 29 **27.** 31 **28.** 47 **29.** 70
30. 11 **31.** 49 **32.** 38 **33.** 22 **34.** 57 **35.** 10.6 **36.** 8.8
37. 9.8 **38.** 3 **39.** 4 **40.** 10.8 **41.** 1 **42.** 14.2 **43.** 12
44. 7.6 **45.** 6 **46.** 4 **47.** 1.3 **48.** 4 **49.** 19 **50.** 5 **51.** 1
52. 3 **53.** 20 **54.** 1 **55.** 0.25 **56.** 10 **57.** 1 **58.** 0.2
59. ⁻0.08 **60.** 0.16

PAGE 122 **1.** $16 \div x$ or $\frac{16}{x}$ **2.** $x + 8$ **3.** $5x$ **4.** $x - 20$
5. $y + 7$ **6.** $y - 9$ **7.** $0.15y$ **8.** $3y + 2$ **9.** $8x + 7$
10. $(x - 6) + 3$ **11.** $(y + 1) - 4$ **12.** $12 + (y - 5)$
13. $34 = x + 17$ **14.** $0.2y = 10$ **15.** $5(4) = x$ **16.** $y \div 3 = 6$
or $\frac{y}{3} = 6$ **17.** $8 + 4 = x$ **18.** $7(3 + y) = 28$ **19.** $3x - 8 = 25$
20. $b = 18 \div 42$ or $b = \frac{18}{42}$ **21.** e **22.** d **23.** b
24. $0.5p = \$65$ **25.** $c = 38 - 9$ **26.** $p = \$40 \div 6$
or $p = \frac{\$40}{6}$ **27.** $c = (3 \times \$5) + \$2$

PAGE 123 **1.** 4 **2.** 11 **3.** 72 **4.** 2 **5.** 9 **6.** 34 **7.** 24
8. 3 **9.** 21 **10.** 12 **11.** 32 **12.** 15 **13.** 6 **14.** 12 **15.** 27
16. 3 **17.** 15 **18.** 3 **19.** 7 **20.** 22 **21.** 30 **22.** 13 **23.** 7
24. 0 **25.** 35 **26.** 52 **27.** 10 **28.** 0 **29.** 21 **30.** 8 **31.** 21
32. 36 **33.** 4 **34.** 20 **35.** 10 **36.** 2 **37.** 20° C **38.** 72° C
39. 175 mi **40.** 4 h **41.** 225 square ft **42.** 60 ft
43. 75 mph **44.** 70 ft

PAGE 124 **1.** 46 **2.** 6.2 **3.** 9 **4.** 48 **5.** 4.1 **6.** 164 **7.** 55
8. 27 **9.** 1.8 **10.** 14 **11.** 5.5 **12.** 8 **13.** 10 **14.** 24 **15.** 27
16. 35 **17.** 7.7 **18.** 19 **19.** 8.5 **20.** 7.2 **21.** 46 **22.** 2.9
23. 8.3 **24.** 74 **25.** 21 **26.** 2 **27.** 11 **28.** ⁻6.6 **29.** 4

30. 1.5 **31.** $p - \$12 = \$78$; $90 **32.** $p + \$6.50 = \$84.20$;
$77.70 **33.** $t - 6 = 88$; 94 points **34.** $w - 7 = 132$;
139 lb **35.** $25\frac{1}{2} + x = 29\frac{3}{4}$; $4\frac{1}{4}$ **36.** $56 - x = 49\frac{3}{4}$; $6\frac{1}{4}$
37. $\$135 + c = \$326$; $191 **38.** $\$895 - e = \$829$; $66
39. $w + 5 = \$162$; 157 lb **40.** $t + 15 = 55$; 40 min
41. $d + 1.5 = 8.0$; 6.5 m **42.** $t + 30 = 75$; 45 min

PAGE 125 **1.** 18 **2.** 75 **3.** 80 **4.** 2 **5.** 14.4 **6.** 20 **7.** 18
8. 160 **9.** 78 **10.** 7 **11.** 12 **12.** 48 **13.** 30 **14.** 60
15. 400 **16.** 40 **17.** 1,600 **18.** 2.05 **19.** 4 **20.** 48 **21.** 8
22. 96 **23.** 9 **24.** 77 **25.** 105 **26.** 48 **27.** 96 **28.** 25
29. 1.25 **30.** 900 **31.** $\frac{n}{6} = 8$; 48 **32.** $5n = 65$; 13
33. $9n = 108$; 12 **34.** $\frac{n}{8} = 7$; 56 **35.** $\frac{n}{7} = 6$; 42
36. $8n = 72$; 9 **37.** $\frac{b}{4} = \$60$; $240 **38.** $0.8p = \$284$;
$355 **39.** $3p = \$5,538$; $1,846 **40.** $\frac{p}{3} = \$745$; $2,234
41. $\frac{b}{3} = \$75$; $225 **42.** $2n = \$84$; $42

PAGE 126 **1.** $13 + 6x = 74$ or $6x + 13 = 74$
2. $5 + 3n = 32$ or $3n + 5 = 32$ **3.** $7 + 2r = 18$ or
$2r + 7 = 18$ **4.** $16 + \frac{t}{3} = 21$ or $\frac{t}{3} + 16 = 21$ **5.** $8 + \frac{p}{4} = 15$
or $\frac{p}{4} + 8 = 15$ **6.** $3 + \frac{z}{5} = 30$ or $\frac{z}{5} + 3 = 30$ **7.** 9 **8.** 8
9. 8 **10.** 32 **11.** 4.3 **12.** 6 **13.** 335 **14.** 70.2 **15.** 192
16. 144 **17.** 6 **18.** 450 **19.** 48 **20.** 30 **21.** 1.5 **22.** 195
23. 154 **24.** 15 **25.** 175 **26.** $2n + 7 = 23$; 8
27. $\frac{n}{6} - 3 = 25$; 168 **28.** $\frac{n}{8} + 7 = 14$; 56
29. $0.6n - 7 = 47$; 90 **30.** $8b + 4 = 100$; 12 pens
31. $\frac{c}{4} - 5 = 4$; 36 boxes **32.** $3d + 45 = 405$; 102 mi
33. $0.9d - 10 = 134$; 160 mi **34.** $\$120 + \frac{s}{6} - \$40 = \$234$;
$924 **35.** $\$110 + 0.18s - \$45 = \$247.70$; $1,015

PAGE 127 **1.** a **2.** a **3.** $187.50 **4.** $9,026.33
5. $8,470 **6.** $11,834.02 **7.** $42,560 **8.** $9,164

RETEACHING SHEETS

PAGE 1 **1.** 9,041 **2.** 3,091 **3.** 9.4 **4.** 7.41 **5.** 10.1
6. 11.24 **7.** 67.21 **8.** 44.21 **9.** 22.1 **10.** 6.25 **11.** 1,122
12. 7,198 **13.** 8,562 **14.** 4,527 **15.** 8,198 **16.** 10,192
17. 6.3 **18.** 7.4 **19.** 22.23 **20.** 48.91 **21.** 22.12
22. 16.24 **23.** 8,203 **24.** 11.09 **25.** 25.25 **26.** 11.02

PAGE 2 **1.** 2,136 **2.** 2,192 **3.** 5.6 **4.** 2.14 **5.** 0.16
6. 4.9 **7.** 2.7 **8.** 3.88 **9.** 5.09 **10.** 8.50 **11.** 4.8 **12.** 72
13. 16.89 **14.** 11.77 **15.** 1,279 **16.** 2.6 **17.** 1.9
18. 1,878 **19.** 5.8 **20.** 7,045 **21.** 7.85 **22.** 2.61 **23.** 6.12
24. 932.1 **25.** 27.93 **26.** 3.3 **27.** 16.51 **28.** 2,577

PAGE 3 **1.** 235 **2.** 4,048 **3.** 27.6 **4.** 3.08 **5.** 17.2 **6.** 5.6
7. 32.58 **8.** 3.78 **9.** 32.24 **10.** 19.36 **11.** 17.5 **12.** 14.58
13. 164 **14.** 318 **15.** 3.65 **16.** 3.6 **17.** 28.64 **18.** 42.21
19. 552 **20.** 83.7 **21.** 2.1 **22.** 4,956 **23.** 75.2 **24.** 1,026
25. 3.33 **26.** 6.06 **27.** 2.8 **28.** 2,020 **29.** 24.36
30. 28.26

PAGE 4 Check students' work for Exercises 1–4.
Quotients are given. **1.** 214 **2.** 543 R3 **3.** 5.6 **4.** 7.43
5. 2.46 **6.** 1.55 **7.** 2.88 **8.** .29 **9.** .71 **10.** 6.25 **11.** 28
12. 23 **13.** 12 R4 **14.** 435 R1 **15.** 171 **16.** 113
17. 50 R6 **18.** 92 **19.** 905 R2 **20.** 1,417 **21.** 2,339
22. 397 R1 **23.** 5.3 **24.** 0.54 **25.** 5.56 **26.** 8.08 **27.** 3.8
28. 4.69

PAGE 5 **1.** 0.28 **2.** 0.57 **3.** 0.13 **4.** 0.08 **5.** 0.035

6. 0.89 **7.** 0.029 **8.** 1.23 **9.** 0.0075 **10.** 2.08 **11.** 0.62
12. 4 **13.** $\frac{7}{20}$ **14.** $\frac{4}{5}$ **15.** $\frac{1}{4}$ **16.** $\frac{2}{5}$ **17.** $\frac{1}{5}$ **18.** $\frac{12}{25}$
19. $\frac{11}{20}$ **20.** $\frac{8}{25}$ **21.** $1\frac{1}{5}$ **22.** $1\frac{3}{4}$ **23.** $1\frac{3}{10}$ **24.** $2\frac{2}{5}$ **25.** 0.4
26. $\frac{3}{100}$ **27.** 12% **28.** $\frac{3}{25}$ **29.** 1.50 **30.** $1\frac{1}{2}$ **31.** 0.075
32. $\frac{3}{40}$

PAGE 6 **1.** 40 **2.** 8 **3.** 24 **4.** 24 **5.** 72 **6.** 11 **7.** 45
8. 72 **9.** 91 **10.** 15 **11.** 96 **12.** 80 **13.** 45 **14.** 55 **15.** 34
16. 16 **17.** $24 **18.** 142 people

PAGE 7 **1.** $126 + 132 + 118 + 122 + 132 = 630$;
$630 \div 5 = 126$; 126 **2.** 126 **3.** 132 **4.** 94 **5.** 14 **6.** 461
7. 38 **8.** 141 **9.** 100 **10.** 14 **11.** 7.3 **12.** 127 **13.** 84
14. 14.9 **15.** 52 **16.** 127 **17.** 282 **18.** 86 **19.** 45
20. 98.6 **21.** 88 **22.** 7

PAGE 8 **1.** 70.55; 86.05 **2.** 268.8; 134.4 **3.** 45.8; 114.5
4. 65.5; 44.06 **5.** 27.12; 41.15; 23.05 **6.** 4.76; 21.52;
5.95 **7.** 339.59 **8.** 202.1 **9.** 99.972 **10.** 1,425
11. 1,145.98 **12.** 1,129.05 **13.** 931.5 **14.** 1,022.26
15. 5.91 **16.** 0.259 **17.** 397.6 **18.** 709 **19.** 35.56
20. 330.603

PAGE 9 **1.** 76; 75 **2.** 467; 465 **3.** 46; 45 **4.** 350; 354
5. 95 **6.** 93 **7.** 663 **8.** 553 **9.** 743 **10.** 85¢ **11.** $342
12. $855 **13.** $6.55 **14.** $6.45 **15.** 21 **16.** 38 **17.** 168
18. 469 **19.** 54¢ **20.** $575 **21.** $318 **22.** $48 **23.** $55.48
24. $12.94

   ANSWER BOOK   Practical Mathematics: Consumer Applications (Practice/Reteaching)

PAGE 10 **1.** 10,000; 58,000 **2.** $30.00; $9.00; $6.00; $45.00 **3.** 3,000; 900; 2,100 **4.** $30.00; $170.00 **5.** 90,000 **6.** 8,000 **7.** 600 **8.** 30 **9.** 3,000 **10.** 6 **11.** 10 **12.** 0.5 **13.** 10 **14.** 0.3 **15.** 40 **16.** 0.9 **17.** $8.00 **18.** 450.00 **19.** 4,690 **20.** $6.00 **21.** 19.90 **22.** 2.00 **23.** 10,000 **24.** $19.20 **25.** 0.65

PAGE 11 **1.** 60 **2.** 100 **3.** 500 **4.** 6,000 **5.** 40,000 **6.** 30 **7.** 7 **8.** 0.8 **9.** 300 **10.** 20 × 200; 4,000 **11.** 300 ÷ 50; 6 **12.** 4 **13.** 16 ÷ 2; 8 **14.** 400 ÷ 2; 200 **15.** 400 **16.** 600 **17.** 30,000 **18.** 30,000 **19.** 32 **20.** $12 **21.** 0.56 **22.** 540 **23.** 20 **24.** 20 **25.** 300 **26.** 3 **27.** 35 **28.** $8 **29.** 5

PAGE 12 **1.** 84 **2.** 30 **3.** 9 **4.** 18 **5.** 36 **6.** 18 **7.** 7 **8.** 2 **9.** 12 **10.** 68 **11.** 43 **12.** 12 **13.** 25; 20; 45 **14.** 48; 18; 30 **15.** 8 × 6; 48 **16.** 27 ÷ 3; 9 **17.** 7 **18.** 9 **19.** 25 **20.** 90 **21.** 105 **22.** 5 **23.** 8 **24.** 55 **25.** 20 **26.** 32 **27.** 54 **28.** 80 **29.** 120 **30.** 8 **31.** 11 **32.** 6 **33.** 9 **34.** 7

PAGE 13 **1.** no **2.** yes **3.** yes **4.** no **5.** no **6.** yes **7.** c **8.** b **9.** c **10.** b **11.** c **12.** a **13.** c **14.** no **15.** yes

PAGE 14 **1.** 6; 13; 78 **2.** 2; 28; 56 **3.** 78; 56; 134 **4.** $66 **5.** $63.50 **6.** $64.75 **7.** $61

PAGE 15 **1.** $100 **2.** $172.50 **3.** $228 **4.** $43.75 **5.** $6.25 **6.** $28 **7.** cashier; $2.50

PAGE 16 **1.** $5 **2.** $3.50 **3.** $4.26 **4.** $1.72 **5.** $8.32 **6.** $6.64 **7.** $7.50 **8.** $5.25 **9.** $6.39 **10.** $2.58 **11.** $12.48 **12.** $9.96 **13.** $10 **14.** $7 **15.** $8.52 **16.** $3.44 **17.** $16.64 **18.** $13.28 **19.** $2.25 **20.** $5.50 **21.** $2.30 **22.** $11.24 **23.** $7.85

PAGE 17 **1.** $37.55 **2.** $60.08 **3.** $105.14 **4.** $56.33 **5.** $24.41 **6.** $47.76 **7.** $71.87 **8.** $9.97 **9.** $28.24 **10.** $59.54 **11.** $55.99 **12.** $86.43 **13.** $33.80 **14.** $416.20 **15.** $29.44 **16.** $362.56 **17.** $43.71 **18.** $538.29 **19.** $96.13 **20.** $1,183.87 **21.** $71.81 **22.** $884.39 **23.** $82.01 **24.** $1,009.99 **25.** $48.68 **26.** $599.57 **27.** $55.33 **28.** $681.36 **29.** $18.14 **30.** $223.36 **31.** $60.62 **32.** $746.63 **33.** $127.84 **34.** $1,574.36

PAGE 18 **1.** c **2.** b **3.** c **4.** b **5.** 30 h; 25 h **6.** 7 quarters; 2 nickels **7.** $45; $70 **8.** June–28 h; July–56 h

PAGE 19 **1.** 4 **2.** 13 **3.** 12 **4.** 6 **5.** 10 **6.** $11.10 **7.** $9.48 **8.** $18.84 **9.** $9.00 **10.** $14.76 **11.** 8; $11.70; $312.00; $93.60; $405.60 **12.** 11; $10.13; $270.00; $111.43; $381.43 **13.** 5; $14.04; $374.40; $70.20; $444.60 **14.** 7; $12.80; $341.20; $89.60; $430.80

PAGE 20 **1.** 11:30 **2.** 1:15 **3.** 9:45 **4.** 3:00 **5.** 10:30 **6.** 9:45 **7.** 6:00 **8.** $8\frac{1}{4}$ h **9.** 9:15 **10.** 5:00 **11.** $7\frac{3}{4}$ h **12.** 8:30 **13.** 4:00 **14.** $7\frac{1}{2}$ h **15.** 9:00 **16.** 5:00 **17.** 8 h **18.** $38\frac{3}{4}$ h **19.** 7:45 **20.** 3:30 **21.** $7\frac{3}{4}$ h **22.** 9:00 **23.** 5:15 **24.** $8\frac{1}{4}$ h **25.** 8:30 **26.** 4:30 **27.** 8 h **28.** 8:45 **29.** 4:15 **30.** $7\frac{1}{2}$ h **31.** 40 h

PAGE 21 **1.** $297.20 **2.** $15,454.40 **3.** $306.60 **4.** $15,943.20 **5.** $370.00 **6.** $19,240.00 **7.** $367.50 **8.** $19,110.00 **9.** 26 **10.** $18,850.00 **11.** 24 **12.** $19,494.00 **13.** 12 **14.** $17,703.84 **15.** 26 **16.** $15,173.08 **17.** Job A; $1,400 **18.** Job A; $1,100 **19.** Job A; $2,028 **20.** Job B; $2,223

PAGE 22 **1.** $18.00 **2.** $112.50 **3.** $105.00 **4.** $48.00 **5.** $281.25 **6.** 400 **7.** $300.00 **8.** 100 **9.** $250.00 **10.** 210 **11.** $262.50 **12.** 160 **13.** $840.00 **14.** $367.50 **15.** $266.00 **16.** $420.00 **17.** $437.50

PAGE 23 **1.** $1,002.63 **2.** 0.08 **3.** $2,046.00 **4.** 0.0825 **5.** $1,020.11 **6.** 0.15 **7.** $4,447.50 **8.** 0.12 **9.** $2,238.36 **10.** 0.0525 **11.** $227.43 **12.** 0.075 **13.** $2,770.65 **14.** 0.18 **15.** $2,677.50 **16.** 0.03 **17.** $1,203.75 **18.** $1,318.60 **19.** $4,587.72 **20.** $1,761.21 **21.** $2,623.61

PAGE 24 **1.** $332.50; $69.88; $262.62 **2.** $185.60; $39.02; $146.58 **3.** $271.70; $57.09; $214.61 **4.** $174.00; $36.54; $137.46 **5.** $305.62; $64.22; $241.40 **6.** $142.20; $29.84; $112.36 **7.** $238.70; $50.17; $188.53 **8.** $184.62; $38.81; $145.81 **9.** $306.80; $64.45; $242.35 **10.** $490.00 **11.** $0.00 **12.** $490 **13.** $53.90 **14.** $34.30 **15.** $9.80 **16.** $98.00 **17.** $392.00

PAGE 25 **1.** 0.8 **2.** $200 **3.** $125 **4.** $1,040 **5.** 0.7 **6.** $728 **7.** $412 **8.** $560 **9.** 0.8 **10.** $448 **11.** $162 **12.** $1,650 **13.** 0.9 **14.** $1,485 **15.** $265 **16.** $2,290 **17.** 0.7 **18.** $1,603 **19.** $762 **20.** $105 **21.** 0.6 **22.** $63 **23.** $302 **24.** $825 **25.** 0.8 **26.** $660 **27.** $265 **28.** $3,570 **29.** 0.7 **30.** $2,499 **31.** $1,171 **32.** $2,060 **33.** 0.8 **34.** $1,648 **35.** $487 **36.** $58.80 **37.** $1,420

PAGE 26 **1.** 30 **2.** $5.74 **3.** $172.20 **4.** 70 **5.** $2.86 **6.** $200.20 **7.** 50 **8.** $28.51 **9.** $1,425.50 **10.** 60 **11.** $18.46 **12.** $1,107.60 **13.** 50 **14.** $15.78 **15.** $789.00 **16.** 25 **17.** $3.90 **18.** $97.50 **19.** 75 **20.** $4.42 **21.** $331.50 **22.** 20 **23.** $23.37 **24.** $467.40 **25.** 55 **26.** $13.67 **27.** $751.85 **28.** $221 **29.** $473.40 **30.** $445.20 **31.** $352.35 **32.** $3,168 **33.** $261

PAGE 27 **1.** $21.25 **2.** $3.75 **3.** $20 **4.** $5 **5.** $16.75 **6.** $3.25 **7.** $13.75 **8.** $6.25 **9.** $20 **10.** $30

PAGE 28 **1.** $21.50 **2.** $16.75 **3.** $22.50 **4.** $16.25 **5.** $25.75 **6.** $25.50 **7.** 4 times **8.** 4 times **9.** $17.50; pay separately **10.** $3 **11.** $4.50 **12.** $3

PAGE 29 **1.** $10.14 **2.** $9.11 **3.** $12.63 **4.** $7.51; $2.49 **5.** $11.14; $3.86 **6.** $27.25; $2.75 **7.** no **8.** yes **9.** yes **10.** no

PAGE 30 **1.** $45.00 **2.** $116.50 **3.** $345.00 **4.** $460.00 **5.** $115.00 **6.** $178.75 **7.** $250.00 **8.** $20.00 **9.** $326.95 **10.** $365.00 **11.** $38.50 **12.** $466.25 **13.** $136.80 **14.** $25.60 **15.** $208.29 **16.** buy racquet **17.** rent ice skates

PAGE 31 **1.** $0.35 **2.** $0.15 **3.** $0.72 **4.** $0.30 **5.** $1.44 **6.** $0.45 **7.** $1.75 **8.** $1.25 **9.** $0.40 **10.** $1.21 **11.** $1.25 **12.** $0.75

PAGE 32 **1.** $27.75 **2.** $157.25 **3.** 20% **4.** $192.20 **5.** $95.23 **6.** $285.67 **7.** 50% **8.** $180.75 **9.** $30.01 **10.** $106.96 **11.** 45% **12.** $62.34

PAGE 33 **1.** $26 **2.** $21.98 **3.** $4.02 **4.** $60 **5.** $44.97 **6.** $15.03 **7.** $34 **8.** $29.90 **9.** $4.10 **10.** $33 **11.** $25.98 **12.** $7.02 **13.** $57 **14.** $44.93 **15.** $12.07 **16.** $73 **17.** $62.87 **18.** $10.13 **19.** $12.03 **20.** $18.14 **21.** $12.07 **22.** $19.13

PAGE 34 **1.** $0.08 **2.** $0.66 **3.** $0.82 **4.** $0.69 **5.** $1.40 **6.** $0.79 **7.** $0.73 **8.** $1.45 **9.** $1.35 **10.** $0.13

11. $0.76 12. $1.37 13. $0.11 14. $0.29 15. $0.97
16. $0.67 17. $4.72 18. $7.88 19. $1.58 20. $3.48
21. $14.75 22. $28.98 23. $13.88 24. $0.92 25. $50.98
26. $154.69 27. $93.04 28. $74.85

PAGE 35 1. $5.25 2. $10.25 3. $7.50 4. $14.75
5. $59.50 6. $74.20 7. $196.15 8. $194.05 9. $57.95
10. $90.98 11. $140.63 12. $120.55 13. $155.51
14. $276.05

PAGE 36 1. 12¢ per oz; 11¢ per oz; 6 oz for 66¢ 2. 48¢
per oz; 44¢ per oz; 16 oz for $7.04 3. 9¢ per oz; 11¢ per
oz; 8 oz for 72¢ 4. 6¢ per oz; 5¢ per oz; 32 oz for $1.60
5. 7¢ per oz; 8.5¢ per oz; 10 oz for 70¢ 6. 9.5¢ per oz;
8.8¢ per oz; 2 lb 8 oz for $3.52 7. $1.60 per lb; $1.80 per
lb; 4.5 lb for $7.20 8. 44¢ per lb; 40¢ per lb; 3.5 lb for
$1.40 9. 76¢ per L; 64¢ per L; 4 L for $2.56 10. $1.60
per lb; $1.65 per lb; 2.75 lb for $4.40 11. 8.5¢ per oz;
7.8¢ per oz; 36 oz for $2.81 12. 40¢ per lb; 42¢ per lb;
4.75 lb for $1.90 13. $43.92 per oz; $34.00 per oz; 0.75
oz for $25.50 14. $7.92 per lb; $7.69 per lb; 6.50 lb for
$50.00

PAGE 37 1. $1 2. $3 3. $1 4. $4 5. $7 6. $3 7. more
than $5 8. less than $5 9. less than $5 10. more than
$5 11. Answers may vary. 12. more than $5

PAGE 38 1. $87.42 2. $498.68 3. $622.09 4. $654.68
5. New balance: $604.50; Check students' work. 6. New
balance: $540.84; Check students' work.

PAGE 39 1. $515.12 2. $497.62 3. 236 4. $23.75
5. $6.25 6. $491.37 7. $421.15 8. $633.57

PAGE 40 1. $115.06 2. $166.28 3. $495.41
4. $220.09 5. $652.93 6. $1,962.98 7. $550.82
8. $515.82 9. $465.82 10. $671.32 11. $847.20
12. $747.20 13. $807.65

PAGE 41 1. $1.1262 2. $1.1248 3. $1.1956
4. $2.0328 5. $1.2184 6. $1.3629 7. $2,343.40
8. $4,031.70 9. $12,867.20 10. $5,710.50
11. $4,388.40 12. $19,663.20 13. $15,265.15
14. $16,624.32

PAGE 42 1. 123 456 789 123 2. 1/15/90 3. 2/17/90
4. $1,500 5. $1,015.87 6. $2.67 7. 19.80% 8. $458.85
9. $484.13 10. $158.31 11. $135.70 12. $35

PAGE 43 1. $8.50 2. $11.00 3. $12.37 4. $12.78
5. $12.97 6. $17.03 7. $7.03 8. $15.46 9. $23.11
10. $450 11. $6.75 12. $706.75 13. $225 14. $3.38
15. $228.38 16. $710 17. $9.60 18. $884.60 19. $385
20. $5.78 21. $853.73 22. $175.60 23. $2.63
24. $393.23 25. $862.75 26. $11.13 27. $1,339.08
28. $562 29. $8.12 30. $570.12 31. $328.84 32. $4.93
33. $498.63 34. $1,000 35. $12.50 36. $1,308

PAGE 44 1. $7.54; $327.04 2. $3.04; $241.54
3. $10.51; $528.16 4. $16.91; $1,246.96 5. $6.68;
$546.68 6. $2.16; $108.16

PAGE 45 1. $131.94 2. $7,916.40 3. $1,916.40
4. $175.92 5. $10,555.20 6. $2,555.20 7. $152.30
8. $18,276.00 9. $8,276.00 10. $208.91 11. $12,534.60
12. $3,034.60 13. $219.21 14. $26,305.20
15. $10,055.20 16. $106.10 17. $6,366 18. $1,541
19. $201.37 20. $36,246.60 21. $20,696.60

PAGE 46 1. $99 2. $19 3. $108 4. $13 5. $135

6. $15 7. $170 8. $12 9. $232.50 10. $17.50 11. $476
12. $20 13. $643 14. $20.50 15. $545.25 16. $49.30
17. $1,011 18. $112.05 19. $1,021.20 20. $64.80
21. $131; $11 22. $953; $58

PAGE 47 1. $12,905 2. $12,190.50 3. $9,392
4. $13,845 5. $9,935 6. $12,472 7. $10,477 8. $9,516
9. $6,916.75 10. $7,111.50 11. $9,200 12. $8,815.25

PAGE 48 1. $1,115.84; $11,915.84 2. $1,487.20;
$11,487.20 3. $2,581; $12,981 4. $2,560.40;
$14,310.40 5. $1,819.20; $13,399.20 6. $3,347.75;
$16,912.60 7. $917.24; $9,812.24 8. $2,781.84;
$16,831.84

PAGE 49 1. $30 2. $40 3. $40 4. $10 5. $10 6. $20
7. $20 8. $10 9. $20 10. $20 11. $60 12. $30

PAGE 50 1. $507.22 2. $477.38 3. $534.73
4. $474.72 5. $478.86 6. $462.31 7. 3.00 8. 1.35
9. 1.30 10. 1.20 11. $1,025.37 12. $882.30

PAGE 51 1. $144.50 2. $229.70 3. $159.55
4. $146.16 5. $273.62 6. $41.00 7. $106.63 8. $150.96
9. $284.82 10. $748.00 11. $982.10 12. $121.62

PAGE 52 1. 5 h 2. 6 h 3. 8.5 h 4. 12.5 h 5. 6.5 h
6. 9.25 h 7. 30 h 8. 40 h 9. 13 h 10. 31 h 11. 30 h
12. 25 h 13. 27 h 14. 42 h 15. 12 h 16. Atlanta to Miami
about 4 h longer

PAGE 53 1. 11:55 A.M. 2. 6:10 P.M. 3. 12:45 P.M.
4. 1:40 P.M. 5. 3 h 35 min 6. 5 h 45 min 7. 8 h 35 min
8. 11 h 15 min 9. 9:27 A.M. 10. 2:17 P.M. 11. 10:43
A.M. 12. 9:59 A.M. 13. 1 h 16 min 14. 2 h 4 min
15. 1 min 16. 18 min

PAGE 54 1. $725 2. $239 3. $950 4. $2,025
5. $1,295 6. $2,200 7. $4,675 8. $1,571 9. $150
10. $336 11. $672 12. $1,025 13. Goldsteins; $614
more 14. Fiona and her sister; $335 less

PAGE 55 1. $1 2. $0.80 3. $0.80 4. $0.80 5. $0.80
6. $2 7. $2.20 8. $1.10 9. $2.05 10. $0.80

PAGE 56 1. $0.90 2. $0.60 3. $1.05 4. $1.40 5. $1.65
6. $1.80 7. $2.35 8. $1.25 9. $6.80 10. $7.95
11. $5.75 12. $9.00 13. $13.50 14. $18.80 15. $24.05
16. $20.25 17. $15.80 18. $24.15

PAGE 57 1. $20 2. $24 3. $20 4. $30 5. $42 6. $53
7. $160 8. $240 9. $240 10. $600 11. $840 12. $720
13. $180 14. $600 15. $630 16. about $370

PAGE 58 1. $11,236.54 2. NASCO Industries
3. $1,714.80 4. $241.62 5. $835.04 6. $85.20 7. Village
Bank 8. $206.44 9. $21,972 10. $3,295.80
11. $1,648.90 12. $472.40 13. $164.79 14. $315.26

PAGE 59 1. $21,375; $16,935 2. $17,095; $12,655
3. $42,380; $34,820 4. $35,592; $26,132 5. $32,760;
$28,320 6. $21,065; $11,605 7. $38,342; $30,782
8. $28,450; $22,110 9. $33,405; $27,065 10. $42,200;
$32,740

PAGE 60 1. 1 2. $15,384 3. 3 4. $15,711.34 5. 7
6. $11,271.34 7. $2,075.55 8. $1,619 9. $456.55

PAGE 61 1. $484; standard 2. $135; itemized 3. $390;
standard 4. $800; standard 5. $518; itemized 6. $72;

PAGE 62   **1.** $419.35 **2.** $767.25 **3.** $673.45
4. $920.73 **5.** $956.60 **6.** $1,154 **7.** $543.43
8. $1,253.58 **9.** $1,047.43 **10.** $658.05

These are correct

PAGE 63   **1.** $1,125 **2.** $2,560 **3.** $675 **4.** $1,451.25
5. $3,200 **6.** $1,069.88 **7.** $1,285.28 **8.** $2,369.44 **9.** the
Briarwood **10.** the Stanton **11.** ~~$3,019.50~~ **12.** ~~$3,290.25~~
$604.75   *$193.50*

PAGE 64   **1.** $6,000 **2.** $66,000 **3.** $3,840 **4.** $51,840
5. $3,750 **6.** $78,750 **7.** $8,515.20 **8.** $79,475.20
9. $8,586 **10.** $103,986 **11.** $5,345.80 **12.** $68,225.80
13. $6,772.13 **14.** $97,067.13 **15.** $66,550 **16.** $96,228
17. $79,615.26 **18.** $68,628.52

PAGE 65   **1.** $12,050; $1,360 **2.** $12,200; $1,663
3. $4,824.25; $1,177 **4.** $7,750; $980 **5.** $21,600;
$1,588 **6.** $9,578.25; $772 **7.** $6,780; $820 **8.** $13,860;
$1,173

PAGE 66   **1.** $519.75 **2.** $571.88 **3.** $811.44
4. $680.40 **5.** $900.59 **6.** $527.30 **7.** $8,750 **8.** $78,750
9. $744.19 **10.** $20,400 **11.** $81,600 **12.** $842.93
13. $10,800 **14.** $61,200 **15.** $621.79 **16.** $7,302.40
17. $83,977.60 **18.** $755.80 **19.** $2,849.50
20. $54,140.50 **21.** $473.19 **22.** $4,503.75
23. $55,546.25 **24.** $564.35

PAGE 67   **1.** $59,500 **2.** $73,600 **3.** $55,500
4. $66,300 **5.** $37,680 **6.** $65,662.50 **7.** $96,064
8. $121,811 **9.** $2,644.10 **10.** $2,338.70 **11.** $980.42
12. $2,677.20 **13.** $4,083.94 **14.** $2,464.20
15. $4,512.44 **16.** $3,860.84 **17.** $86,400; $3,654.72
18. $67,087.50; $3,401.34

PAGE 68   **1.** $6,800 **2.** $34,000 **3.** $13,600 **4.** $3,400
5. $5,200 **6.** $26,000 **7.** $10,400 **8.** $2,600 **9.** $7,900
10. $39,500 **11.** $15,800 **12.** $3,950 **13.** $8,250
14. $41,250 **15.** $16,500 **16.** $4,125 **17.** $12,200
18. $61,000 **19.** $24,400 **20.** $6,100 **21.** $9,280
22. $46,400 **23.** $18,560 **24.** $4,640 **25.** $10,720
26. $53,600 **27.** $21,440 **28.** $5,360 **29.** $8,285
30. $41,425 **31.** $16,570 **32.** $4,142.50 **33.** $6,124
34. $30,620 **35.** $12,248 **36.** $3,062 **37.** $9,662.50
38. $48,312.50 **39.** $19,325 **40.** $4,831.24 **41.** $11,850
42. $59,250 **43.** $23,700 **44.** $5,925 **45.** $4,926
46. $24,630 **47.** $9,852 **48.** $2,463 **49.** $37,125
50. $24,140

PAGE 69   **1.** 1,466 **2.** $62.74 **3.** 3,057 **4.** $157.74
5. 6,349 **6.** $396.18 **7.** 7,194 **8.** $377.69 **9.** 11,111
10. $684.44 **11.** 16,766 **12.** $648.84 **13.** 13,434
14. $381.53 **15.** $175.85 **16.** $341.06 **17.** $359.65
18. $139.11

PAGE 70   **1.** 8 in.; 6 in. **2.** 8 in.; 6 in. **3.** 36 cm; 24 cm
4. 10 in.; 8 in. **5.** 21 cm; 18 cm **6.** 10 in.; 6 in. **7.** 14 in.;
12 in. **8.** 40 in.; 35 in. **9.** 18 in.; 15 in. **10.** 14 cm; 8 cm
11. 14 in.; 9 in. **12.** Check students' drawings. The scale
drawing should be a rectangle that is 10 cm long and 8
cm wide.

PAGE 71   **1.** 5 **2.** 4 **3.** 4 **4.** 6 **5.** 9 **6.** 10 **7.** 6 **8.** 11
9. $19\frac{1}{2}$ **10.** 45 **11.** 40 **12.** 30 **13.** 40 **14.** 30 **15.** 204
16. 120 **17.** 216 **18.** 144 **19.** 60 **20.** 336 **21.** 360
22. 1,050

PAGE 72   **1.** $165.15 **2.** $215.17 **3.** $1,230.30
4. $1,034 **5.** $173.12 **6.** $488 **7.** $1,058.92
8. $1,150.50 **9.** $602.45 **10.** $457.24

PAGE 73   **1.** $280 **2.** $609 **3.** $787.50 **4.** $1,256
5. $2,208 **6.** $2,320 **7.** $1,708.50 **8.** $4,200
9. $2,092.50 **10.** $6,240 **11.** $661.50 **12.** $935
13. $1,092.50 **14.** $917.50

PAGE 74   **1.** 96 square ft **2.** 51 square ft **3.** 45 square
ft **4.** 1 **5.** $14 **6.** 180 square ft **7.** 51 square ft **8.** 129
square ft **9.** 1 **10.** $14 **11.** 135 square ft **12.** 51 square
ft **13.** 84 square ft **14.** 1 **15.** $14 **16.** 360 square ft
17. 51 square ft **18.** 309 square ft **19.** 2 **20.** $28 **21.** 216
square ft **22.** 102 square ft **23.** 114 square ft **24.** 1
25. $14 **26.** 3 gal; $46.50 **27.** 2 gal; $29.50

PAGE 75   **1.** 3 **2.** $20.25 **3.** 6 **4.** $40.50 **5.** 17
6. $114.75 **7.** 20 **8.** $135 **9.** 31 **10.** $209.25 **11.** 58
12. $391.50 **13.** 121 **14.** $816.75 **15.** 248 **16.** $1,674
17. 287 **18.** $1,937.25 **19.** 325 **20.** $2,193.75 **21.** 472
22. $3,186 **23.** 502 **24.** $3,388.50 **25.** 12 bags; $90
26. 34 bags; $236.30

PAGE 76   **1.** 16 **2.** $284 **3.** 13 **4.** $230.75 **5.** 13.5
6. $239.63 **7.** 50 **8.** $887.50 **9.** 40 **10.** $710 **11.** 25
12. $443.75 **13.** 32.25 **14.** $572.44 **15.** 31.5
16. $559.13 **17.** 11.25 **18.** $199.69 **19.** 140 **20.** $2,485
21. $1,476 **22.** $3,731.50 **23.** 72; $1,212.48 **24.** 24;
$222.24

PAGE 77   **1.** Weeks 3, 4, and 5 **2.** Week 6 **3.** Car A; 150
more **4.** Wolf **5.** Atlas **6.** June, August, September, and
November **7.** 5,000 fewer **8.** Atlas **9.** Brute

PAGE 78   **1.** 3,888 **2.** 91,968 **3.** 1,162 **4.** 396 **5.** 860
6. 52 **7.** 243 **8.** 4,023 **9.** 288 **10.** 39 **11.** 21 **12.** 9,340
13. 455 **14.** 273 **15.** $38.69 **16.** $82.29 **17.** $967.74
18. $86.99 **19.** $126.58 **20.** $15.14

PAGE 79   **1.** 270 **2.** 17 **3.** 192 **4.** 240 **5.** 360 **6.** 90
7. 340 **8.** 850 **9.** 450 **10.** 8,250 **11.** 1,800 **12.** 120
items **13.** 2,250 mufflers **14.** 240 sets **15.** 64 radios

PAGE 80   **1.** 31 drops per min **2.** 35 drops per min **3.** 28
drops per min **4.** 42 drops per min **5.** 52 drops per min
6. 83 drops per min **7.** 240 mL each h **8.** 80 mL each h
9. 150 mL each h **10.** 30 mL each h

PAGE 81   **1.** 18 min 15 s **2.** 19 min 23 s **3.** 17 min 15 s
4. 11 min 45 s **5.** 10 min 37 s **6.** 12 min 45 s **7.** 13 min
27 s **8.** 16 min 10 s **9.** 13 min 13 s **10.** 16 min 15 s

PAGE 82   **1.** 10.1% **2.** 6.7% **3.** 13.6% **4.** 16.7%
5. 7.8% **6.** 18.8% **7.** 20% **8.** 14.5% **9.** 6.2%
10. 10.6% **11.** 13.3% **12.** 13.8% **13.** 6.3% **14.** 11.9%
15. 7.3% **16.** 15.4% **17.** 13.8% **18.** 2.6%

PAGE 83   **1.** $160 **2.** $109 **3.** $66 **4.** $40 **5.** $71
6. $37 **7.** $13 **8.** $19 **9.** $42 **10.** $64 **11.** $62 **12.** $39
13. $44 **14.** $21 **15.** $309 **16.** $116 **17.** $97 **18.** $74

PAGE 84   **1.** $46 **2.** $20 **3.** $275 **4.** $15 **5.** $192
6. $121 **7.** $58 **8.** $25 **9.** $86.75 **10.** $170 **11.** $325
12. $125 **13.** $141 **14.** $102 **15.** $733

PAGE 85   **1.** $29,500 **2.** $31,560 **3.** $33,640
4. $265,000 **5.** $283,500 **6.** $312,600 **7.** $135,680
8. $58,470 **9.** $450,740 **10.** $406,880 **11.** $6,000

itemized **7.** $434; itemized **8.** $1,138; standard **9.** $290;
itemized **10.** $120; standard **11.** $8; standard **12.** $183;
itemized

12. $22,000 **13.** $30,000 **14.** $15,000 **15.** $23,000
16. $12,000

PAGE 86 **1.** $4,278 **2.** $29,600 **3.** $7,990 **4.** $2,620
5. $77,800 **6.** $92,453 **7.** $34,244 **8.** $58,209

PAGE 87 **1.** $121.16 **2.** $233.34 **3.** $329.20 **4.** $68.88
5. $434.30 **6.** $578.24 **7.** $806.80 **8.** $4,186.00
9. $47.31 **10.** $135.18 **11.** $206.64 **12.** $56.64
13. $388.90 **14.** $138.90 **15.** $484.64 **16.** $84.64
17. $806.80 **18.** $306.80 **19.** $369.60 **20.** $19.60

PAGE 88 **1.** 7.92% **2.** 7.54% **3.** 7.38% **4.** 8.03%
5. $79.20 **6.** $237.60 **7.** $118.80 **8.** $188.50 **9.** $5
10. $830 **11.** $184.50 **12.** $37.50 **13.** $94.25
14. $633.60 **15.** $2,409 **16.** $276.75 **17.** $754
18. $187.60 **19.** $746 **20.** $1,993.75

PAGE 89 **1.** 94 **2.** $12\frac{1}{2}$ **3.** 108,100 **4.** 7,100 **5.** 1.12
6. 0.22 **7.** $4\frac{1}{4}$ **8.** $36\frac{3}{4}$ **9.** Edmond **10.** Equitax **11.** Exet
Rf **12.** EnwSy **13.** EGadw **14.** $3\frac{1}{4}$ **15.** $5\frac{3}{4}$ **16.** 3 **17.** $1\frac{5}{8}$
18. 6

PAGE 90 **1.** Daley Products **2.** State of Utah **3.** Western
Metals **4.** City of Prescott **5.** Daley Products, State of
Utah, Western Metals **6.** Kaplan International, City of
Prescott **7.** $330 **8.** $487.50 **9.** $3,687.50
10. $1,437.50 **11.** $6,500 **12.** $7,000

PAGE 91 **1.** $10.74 **2.** $537 **3.** $2,685 **4.** $16.45
5. $822.50 **6.** $4,112.50 **7.** $9.56 **8.** $478 **9.** $2,390
10. $7.85 **11.** $392.50 **12.** $1,962.50 **13.** $24.76
14. $1,238 **15.** $6,190 **16.** $11.01 **17.** $550.50
18. $2,752.50 **19.** $14.12 **20.** $706 **21.** $3,530
22. $7,308 **23.** $14,603 **24.** $838.80 **25.** $18,090
26. $32,900 **27.** $22,944 **28.** $92 **29.** $71.50

PAGE 92 **1.** 1.67 **2.** $7,815.60 **3.** $651.30 **4.** 1.92
5. $7,104 **6.** $592 **7.** 1.37 **8.** $6,165 **9.** $513.75
10. 1.48 **11.** $3,987.12 **12.** $332.36 **13.** 1.95
14. $16,289.52 **15.** $1,357.46 **16.** 2.0 **17.** $8,389.50
18. $699.13 **19.** 2.0 **20.** $10,170 **21.** $847.50 **22.** 2.0
23. $19,188.40 **24.** $1,599.03 **25.** 1.93 **26.** $13,891.37
27. $1,157.61 **28.** 1.98 **29.** $8,511.62 **30.** $709.30
31. $14,576.25; $1,214.69 **32.** $23,234.80; $1,936.23
33. $9,905.70; $825.48 **34.** $20,952.40; $1,746.03

PAGE 93 **1.** $\frac{1}{6}$ **2.** $\frac{1}{6}$ **3.** $\frac{1}{6}$ **4.** $\frac{1}{2}$ **5.** $\frac{1}{3}$ **6.** $\frac{5}{6}$ **7.** $\frac{1}{6}$ **8.** $\frac{1}{6}$
9. $\frac{1}{6}$ **10.** $\frac{5}{6}$ **11.** $\frac{5}{6}$ **12.** $\frac{1}{3}$ **13.** $\frac{1}{2}$ **14.** $\frac{2}{3}$ **15.** $\frac{1}{15}$ **16.** $\frac{2}{15}$
17. $\frac{1}{5}$ **18.** $\frac{7}{15}$ **19.** $\frac{8}{15}$ **20.** $\frac{3}{5}$ **21.** $\frac{1}{3}$ **22.** 0 **23.** $\frac{1}{3}$ **24.** $\frac{1}{15}$
25. $\frac{1}{52}$ **26.** $\frac{1}{26}$ **27.** $\frac{1}{13}$ **28.** $\frac{1}{4}$ **29.** $\frac{10}{13}$ **30.** $\frac{3}{4}$ **31.** $\frac{12}{13}$
32. 0

PAGE 94 **1.** (B,B,B) **2.** (B,B,G) **3.** (B,G,B) **4.** (B,G,G)
5. (G,B,B) **6.** (G,B,G) **7.** (G,G,B) **8.** (G,G,G) **9.** (H,H),
(H,T), (T,H), (T,T) **10.** (H,H,H), (H,H,T), (H,T,H),
(H,T,T),(T,H,H), (T,H,T), (T,T,H), (T,T,T) **11.** (H,H,H,H),
(H,H,H,T), (H,H,T,H), (H,H,T,T), (H,T,H,H), (H,T,H,T),
(H,T,T,H), (H,T,T,T), (T,H,H,H), (T,H,H,T), (T,H,T,H),

(T,H,T,T), (T,T,H,H), (T,T,H,T), (T,T,T,H), (T,T,T,T)
12. (H,1), (H,2), (H,3), (H,4), (H,5), (H,6), (T,1), (T,2), (T,3),
(T,4), (T,5), (T,6) **13.** (L,A), (L,B), (L,C), (R,A), (R,B), (R,C)
14. (1,X), (1,Y), (1,Z), (2,X), (2,Y), (2,Z), (3,X), (3,Y), (3,Z)
15. (1,1), (1,2), (1,3), (1,4), (2,1), (2,2), (2,3), (2,4), (3,1),
(3,2), (3,3), (3,4), (4,1), (4,2), (4,3), (4,4) **16.** (A,1), (A,2),
(A,3), (A,4), (A,5), (A,6), (B,1), (B,2), (B,3), (B,4), (B,5),
(B,6), (C,1), (C,2), (C,3), (C,4), (C,5), (C,6)

PAGE 95 **1.** $\frac{1}{36}$ **2.** 0 **3.** $\frac{1}{4}$ **4.** $\frac{1}{6}$ **5.** $\frac{5}{36}$ **6.** $\frac{1}{6}$ **7.** $\frac{1}{16}$
8. $\frac{3}{32}$ **9.** $\frac{9}{64}$ **10.** $\frac{15}{64}$ **11.** $\frac{15}{64}$ **12.** $\frac{15}{32}$ **13.** $\frac{1}{12}$ **14.** $\frac{1}{6}$
15. $\frac{1}{4}$ **16.** $\frac{1}{3}$ **17.** $\frac{1}{4}$ **18.** $\frac{1}{6}$

PAGE 96 **1.** $\frac{1}{20}$ **2.** $\frac{1}{10}$ **3.** $\frac{3}{10}$ **4.** $\frac{3}{10}$ **5.** $\frac{3}{10}$ **6.** 0 **7.** $\frac{2}{15}$
8. $\frac{1}{10}$ **9.** $\frac{1}{15}$ **10.** $\frac{7}{30}$ **11.** $\frac{7}{15}$ **12.** $\frac{4}{15}$ **13.** $\frac{1}{132}$ **14.** $\frac{5}{22}$
15. $\frac{3}{11}$ **16.** 0 **17.** $\frac{1}{22}$ **18.** $\frac{1}{22}$

PAGE 97 **1.** 10 **2.** 31 **3.** 16 **4.** 26 **5.** 365 **6.** 73
7. 30% **8.** 0.263 **9.** 6 **10.** 3 **11.** 5 **12.** 11 **13.** 20 **14.** 9

PAGE 98 **1.** 7 **2.** 17 **3.** 11 **4.** 8 **5.** 5 **6.** 26 **7.** 10 **8.** 12
9. 33 **10.** 20 **11.** 10 **12.** 20 **13.** 59 **14.** 16 **15.** 35
16. 19 **17.** 40 **18.** 9 **19.** 36 **20.** 147 **21.** 16 **22.** 22
23. 2 **24.** 4 **25.** 7 **26.** 8 **27.** 6 **28.** 1 **29.** 0.5 **30.** 8
31. 85 **32.** 25

PAGE 99 **1.** $x + 4$ **2.** $y - 4$ **3.** $w - 12$ **4.** $\frac{y}{3}$ **5.** $n + 7$
6. $8 + w$ **7.** $5x$ **8.** $16 - y$ **9.** $7n$ **10.** $\frac{r}{10}$ **11.** $0.25x$ or $\frac{x}{4}$
12. $4 - n$ **13.** $6y + 3$ **14.** $12w - 9$ **15.** $7w - 4$ **16.** $\frac{n}{4} + 6$
17. $5w - 4$ **18.** $9n + 12$ **19.** $8 - 5w$ **20.** $\frac{x}{3} + 9$
21. $y + 6 = 10$ **22.** $8 \cdot 3 = w$ **23.** $20 = n + 9$ **24.** $\frac{y}{4} = 2$
25. $x - 9 = 7$ **26.** $4(y + 2) = 12$ **27.** $x - 7 = 25$
28. $2w - 3 = 5$

PAGE 100 **1.** 13 **2.** 4 **3.** 72 **4.** 2 **5.** 17 **6.** 6 **7.** 3 **8.** 16
9. 48 **10.** 14 **11.** 34 **12.** 40 **13.** 7 **14.** 3 **15.** 10 **16.** 2.5
17. 9 **18.** 13 **19.** 16 **20.** 5 **21.** 7.5 **22.** 7 **23.** 5 **24.** 35
25. 50°F **26.** 25°C **27.** 175 mi **28.** 196 square ft

PAGE 101 **1.** 21 **2.** 9.1 **3.** 5.3 **4.** 8.6 **5.** 248
6. 5 **7.** 18.1 **8.** 12 **9.** 2.3 **10.** 24 **11.** 3.2 **12.** 8.1
13. 9.7 **14.** 3.3 **15.** 12.5 **16.** $x - \$4 = 32$; $36
17. $x + \$7.50 = \$21$; $23.50 **18.** $x + 9 = 92$; 83
19. $\$1,488 - x = \$1,306$; $182 **20.** $\$156 + x = \$243$;
$87 **21.** $x + \$105 = \$322$; $217 **22.** $18 - x = 13\frac{1}{2}$;
$4\frac{1}{2}$ **23.** $22\frac{1}{2} + x = 28\frac{1}{4}$; $5\frac{3}{4}$

PAGE 102 **1.** 32 **2.** 164 **3.** 222 **4.** 32.5 **5.** 60 **6.** 25
7. 12 **8.** 3.2 **9.** 112 **10.** 2.5 **11.** 20 **12.** 3 **13.** 5 **14.** 7
15. 126 **16.** 26.4 **17.** 59.5 **18.** 4 **19.** $\frac{1}{4}n = 9$; 36
20. $9n = 72$; 8 **21.** $\$382 = 0.8n$; $477.50 **22.** $3n = \$840$;
$280 **23.** $\$3,230 = 5n$; $646 **24.** $\$2,265 = \frac{n}{2}$; $4,530
25. $\frac{n}{4} = \$42$; $168 **26.** $\$180 \div \$7.50 = n$; 24 h

PAGE 103 **1.** 5 **2.** 3 **3.** 7 **4.** 3.5 **5.** 5 **6.** 10 **7.** 24
8. 12 **9.** 2 **10.** 90 **11.** 30 **12.** 144 **13.** 1.4 **14.** 81
15. 60 **16.** 2 **17.** 45 **18.** 24 **19.** $\frac{2}{3}n - 5 = 11$; 24
20. $5n - 3 = 0.5$; 0.7 **21.** $8n + 3 = 51$; 6 tapes
22. $0.5n - 4 = 20$; 48 bottles **23.** $4n + 4 = 22$; 4.5 mi
24. $\$160 + 0.25s = \$235$; $300 **25.** $30 = 0.75n - 6$;
48 packages **26.** $\$200 + 0.08n = \$440$; $3,000

APPLICATIONS SHEETS

PAGE 1 **1.** Check students' graphs. **2.** Check students'
graphs.

PAGE 2 **1.** 25; 2.00; 2.00; 25; 2.25 **2.** 5.00; 40.00; 40.00;
5.00; 45.00 **3.** 2.00; 25; 50.00; 50.00; 2.00; 25; 51.75

   ANSWER BOOK Practical Mathematics: Consumer Applications (Reteaching/Applications)

4. 5.00; 30; 10.00; 10.00; 5.00; 30; 14.70 5. 2.00; 50;
20.00; 20.00; 2.00; 50; 17.50

PAGE 3 1. $60; $40; $100 2. $190; $170; $170 3. $6\frac{1}{4}$h;
$3\frac{3}{4}$ h; $7\frac{1}{2}$ h 4. $5\frac{1}{4}$ h; 4 h; $4\frac{1}{2}$ h 5. 45 min; 50 min;
30 min 6. $4,800; $1,200; $600

PAGE 4 1. 39.25 2. 0 3. $229.61 4. $0 5. $229.61
6. 30.25 7. 0 8. $176.96 9. $0 10. $176.96 11. 48.5
12. 8.5 13. $234.00 14. $74.63 15. $308.63 16. 45.5
17. 5.5 18. $234.00 19. $48.29 20. $282.29 21. 41.0
22. 1.0 23. $234.00 24. $8.78 25. $242.78 26. 40.5
27. 0.5 28. $234.00 29. $4.39 30. $238.39

PAGE 5 1. $7.53 2. $6.30 3. $28.67 4. $14.94 5. The
membership fee is too high for a short-term class.
6. 5 × $15 = $75; 5 × $174 = $870; $945
7. 2 × $470 = $940; $1,590 8. $945 ÷ 150 = $6.30
9. $1,590 ÷ 150 = $10.60 10. Fitness Health Club
11. Community Center

PAGE 6 1. $23.32 2. $6.63 3. $31.87 4. $24.08
5. $133.50 6. $106.50 7. $76.50 8. $103.50 9. It is the
least expensive. 10. Since tailoring is their only business,
the quality of the work might be better and worth the extra
money.

PAGE 7 1. $211.99 2. $8.11 3. $22.65 4. $810.00
5. $211.99 6. $8.11 7. $22.66 8. $810.00 9. Both
arrangements work out the same way.

PAGE 8 1. $1,950 2. yes 3. $1\frac{1}{2}$% 4. $\frac{1}{2}$% 5. 2%
6. Call the credit-union computer to find out if his last
deposit had cleared yet. 7. Call the credit-union computer
to find out her checking balance. If it is not enough,
transfer some money from her savings account into her
checking account.

PAGE 9 1. $5.38; $440.38; $66.06 2. $12.72; $897.72;
$107.73 3. $9.73; $649.48; $32.47 4. $18.04; $1,568.89;
$235.33

PAGE 10 1. $8.59; $7.94; $7.16 2. $14.39; $13.14;
$11.64; $9.89 3. $20.35; $17.45; $13.97; $9.91; $5.27

PAGE 11 1. 0.26; 6.50; 6.50; 0.05 2. 0.41; 143.50;
143.50; 0.22 3. $0.25 per mi 4. $0.20 per mi 5. $0.14
per mi 6. $0.08 per mi 7. The cars and total distances
are the same. As the number of days (and free miles)
increases, the costs per mile decrease.

PAGE 12 1. $0.26 2. $0.34 3. $0.10 4. $0.04 5. $0.20
6. $0.24 7. $0.09 8. $0.04 9. $0.19 10. $0.23

11. $0.08 12. $0.04 13. longer flights 14. driving

PAGE 13 1. $2,795; $585; $1,625; $698.75; $146.25
2. $1,886.50; $392; $1,225; $471.63; $98 3. $3,652;
$766.92; $2,200; $913; $191.73 4. $3,640; $910; $2,800;
$910; $227.50 5. $6,200; $1,767; $4,650; $1,550;
$441.75 6. $5,062.50; $1,037.81; $3,375; $1,265.63;
$259.45

PAGE 14 1. $15.31 + $15.16 + $1.83 + $0.92 = $33.22
2. $2.59; $7.14; $96.22 3. $1.82; $5.01; $67.58 4. $0.87;
$2.40; $32.41 5. $3.69; $10.15; $136.86

PAGE 15 1. 62 km 2. 29 km 3. 17; 28 4. 1.5 m; 75
cm 5. 4.6 m or 460 cm

PAGE 16 1. billboards, signs, lighted displays, etc.
2. B 3. D 4. B 5. B 6. C 7. D 8. D 9. C 10. A and C;
Household products and fast food are products for the
general population. 11. B; Farmers are interested in
trucks, tractors, and other equipment. 12. $990,000
13. $4,740,000 14. $0 15. $4,005,000 16. $2,685,000
17. $1,365,000 18. $1,080,000 19. $135,000

PAGE 17 1. 1 h 18 min 2. 2 h 36 min 3. 134 h 4. 52
wk, or 1 y 5. 42 min 6. 1 h 24 min 7. 60 h 8. 46 wk
9. 2.31 h, or 2 h 18.6 min 10. 33.53 h, or 33 h 31.8 min
11. Answers may vary.

PAGE 18 1. 192 bonds 2. 139 bonds; $3,475 3. 48
bonds; $1,200 4. 80 bonds; 9.5 y 5. 400 bonds; 6.0 y
6. 300 bonds; 3.0 y 7. 5,000 bonds; 9.5 y

PAGE 19 1. combinations; 10; 1, 2, 3; 1, 2, 4; 1, 2, 5; 1,
3, 4; 1, 3, 5; 1, 4, 5; 2, 3, 4; 2, 3, 5; 2, 4, 5; 3, 4, 5
2. arrangements; 60; 1, 2, 3; 1, 2, 4; 1, 2, 5; 1, 3, 2; 1, 3,
4; 1, 3, 5; 1, 4, 2; 1, 4, 3; 1, 4, 5; 1, 5, 2; 1, 5, 3; 1, 5, 4; 2,
1, 3; 2, 1, 4; 2, 1, 5; 2, 3, 1; 2, 3, 4; 2, 3, 5; 2, 4, 1; 2, 4, 3;
2, 4, 5; 2, 5, 1; 2, 5, 3; 2, 5, 4; 3, 1, 2; 3, 1, 4; 3, 1, 5; 3, 2,
1; 3, 2, 4; 3, 2, 5; 3, 4, 1; 3, 4, 2; 3, 4, 5; 3, 5, 1; 3, 5, 2; 3,
5, 4; 4, 1, 2; 4, 1, 3; 4, 1, 5; 4, 2, 1; 4, 2, 3; 4, 2, 5; 4, 3, 1;
4, 3, 2; 4, 3, 5; 4, 5, 1; 4, 5, 2; 4, 5, 3; 5, 1, 2; 5, 1, 3; 5, 1,
4; 5, 2, 1; 5, 2, 3; 5, 2, 4; 5, 3, 1; 5, 3, 2; 5, 3, 4; 5, 4, 1; 5,
4, 2; 5, 4, 3 3. combinations; 10; 1, 2; 1, 3; 1, 4; 1, 5; 2, 3;
2, 4; 2, 5; 3, 4; 3, 5; 4, 5 4. combinations; 15; 1, 2, 3, 4; 1,
2, 3, 5; 1, 2, 3, 6; 1, 2, 4, 5; 1, 2, 4, 6; 1, 2, 5, 6; 1, 3, 4, 5;
1, 3, 4, 6; 1, 3, 5, 6; 1, 4, 5, 6; 2, 3, 4, 5; 2, 3, 4, 6; 2, 3, 5,
6; 2, 4, 5, 6; 3, 4, 5, 6

PAGE 20 1. 4.9 million 2. 3.2 million 3. 8.2 million
4. 8.3 million 5. 7.1 million 6. 3.2 million 7. 3.0 million
8. 2.9 million 9. 4.2 million 10. 3.9 million 11. 0.9
million 12. 8.0 million 13. 5.3 million 14. 29.2 million

GROUP PROJECTS

Solutions will vary because students are required to
formulate problems and generate their own data. For
each project, encourage students to apply the problem-
solving skills and strategies taught in the related chapter
and preceding chapters.

51 questions

TESTS

Skills Inventory

Part A 1. d 2. b 3. b 4. a 5. b 6. d 7. a 8. b 9. c
10. a 11. d 12. d 13. c 14. a 15. b 16. c 17. b 18. b
19. b 20. a 21. b 22. a 23. d **Part B** 24. c 25. d
26. a 27. b 28. b 29. a 30. b 31. c 32. b 33. c 34. d
35. a 36. a 37. a 38. a 39. c 40. b 41. a 42. b 43. d

44. a 45. a 46. c 47. b 48. d **Part C** 49. a 50. a
51. b 52. c 53. c 54. a 55. a 56. a 57. c 58. b 59. a
60. b 61. c 62. d 63. a 64. d 65. c 66. a 67. b 68. c
69. a 70. d 71. b 72. b 73. a 74. b 75. c 76. b 77. a
78. d 79. a 80. b

CHAPTERS 1–4 1. c 2. b 3. b 4. a 5. c 6. a 7. d
8. a 9. d 10. a 11. a 12. c 13. b 14. d 15. d 16. c
17. b 18. a 19. d 20. b 21. c 22. a 23. a 24. b 25. a
26. d 27. c 28. d 29. b 30. b 31. c 32. d 33. c 34. c
35. a 36. d 37. b 38. c 39. c 40. a

CHAPTERS 1–8 1. c 2. d 3. b 4. d 5. a 6. d 7. c
8. c 9. b 10. b 11. a 12. b 13. c 14. a 15. c 16. d
17. b 18. b 19. c 20. a 21. c 22. a 23. a 24. d 25. b
26. d 27. a 28. c 29. c 30. b 31. d 32. b 33. b 34. a
35. b 36. b 37. c 38. d 39. a 40. b

CHAPTERS 1–12 1. b 2. a 3. c 4. a 5. b 6. c 7. a
8. b 9. b 10. c 11. d 12. a 13. b 14. c 15. a 16. b
17. a 18. d 19. b 20. d 21. a 22. b 23. c 24. a 25. b
26. b 27. a 28. d 29. c 30. b 31. a 32. b 33. d 34. b
35. d 36. c 37. b 38. b 39. c 40. a

CHAPTERS 1–16 1. c 2. b 3. a 4. a 5. c 6. a 7. a
8. c 9. c 10. b 11. b 12. d 13. d 14. a 15. a 16. c
17. b 18. b 19. b 20. c 21. a 22. c 23. a 24. a 25. b
26. b 27. a 28. b 29. b 30. a 31. a 32. c 33. d 34. b
35. b 36. c 37. b 38. b 39. a 40. b

Posttests

CHAPTER 1, Form A 1. 10,917 2. 8,736 3. 49.2
4. 88.2 5. 2,024 6. 4,784 7. 17.6 8. 21.76 9. 360
10. 1,435 11. 22.2 12. 1.26 13. 245 14. 762 15. 2.07
16. 1.36 17. 32% 18. 440% 19. 2.42 20. $1\frac{1}{5}$ 21. 90
22. 1.2 23. 4 24. 110 25. 55 26. 6.7 27. 70.5 28. 6.8
29. Nancy 30. 136 sold 31. 325 people 32. 100 more
attended the dance.

CHAPTER 1, Form B 1. 6,263 2. 10,322 3. 36.3
4. 72.44 5. 2,393 6. 2,786 7. 15.6 8. 21.75 9. 272
10. 4,248 11. 17.2 12. 3.6 13. 451 14. 652 15. 7.05
16. 1.5 17. 45% 18. 380% 19. 7.16 20. $2\frac{3}{5}$ 21. 126
22. 2.7 23. 18 24. 180 25. 2.9 26. 4.7 27. 63.5
28. 6.2 29. Tina 30. 3 more 31. 65 books 32. Jane

CHAPTER 2, Form A 1. 2,549.12 2. 7,771.828
3. 232.05 4. 23.88 5. 767 6. 25¢ 7. 25,000 8. 0.0263
9. $10 10. $10 11. 5.2 12. 9 13. 100,000 14. 180,000
15. 8 16. 7 17. b 18. c 19. c 20. a 21. d 22. $8.33
23. c 24. $45.96 25. a 26. $2.05 27. b 28. $8\frac{1}{8}$ ft
29. no 30. 10 mi 31. $80 32. 20 mi per gal

CHAPTER 2, Form B 1. 4,647.872 2. 672.75
3. 250,464 4. 87.845 5. 4.45 6. $3.44 7. 520 8. 30.52
9. 14 10. 8 11. 470 12. 780 13. 1,800 14. 1,800
15. $5 16. $50 17. c 18. b 19. b 20. a 21. a 22. 62
problems 23. c 24. $34,553.75 25. d 26. $2.75 27. b
28. $2\frac{5}{8}$ ft 29. 21 h 30. 16 tape recorders 31. 160 mi
32. yes

CHAPTER 3, Form A 1. $53 2. $65.87 3. $47.40
4. $45.60 5. $4.05 6. $3.72 7. $4.98 8. $3.90 9. $4.40
10. $1.90 11. $7.90 12. $4.10 13. $47.65 14. $71.95
15. $45.35 16. $90.10 17. $25.50; $1.92 18. $22; $1.65
19. $76.20; $5.72 20. $57.10; $4.29 21. 30 h; 45 h
22. 7 quarters, 3 dimes, 6 nickels 23. $6\frac{1}{2}$ h Monday;
$7\frac{1}{4}$ h Tuesday 24. $22; $37

CHAPTER 3, Form B 1. $61 2. $84.08 3. $34
4. $29.25 5. $4.80 6. $3.75 7. $4.25 8. $4.33 9. $3.60
10. $5.80 11. $6.50 12. $5.10 13. $68.45 14. $52.85
15. $39.65 16. $76.95 17. $22; $1.65 18. $19.50; $1.46

19. $63.20; $4.75 20. $114.75; $8.62 21. 32 h; 46 h
22. 3 quarters, 2 dimes, 12 nickels 23. $5\frac{3}{4}$ h Tuesday;
4 h Thursday 24. $55; $31

CHAPTER 4, Form A 1. $218.50 2. $180.08
3. $298.15 4. $262.96 5. $172.13 6. $183.96 7. $48.26
8. $57.60 9. $157.50 10. 225 pages 11. $83.70
12. 23 h 13. $237.50 14. $189.00 15. $127.84
16. $298.27 17. $11,725 18. $14,885 19. $662.50
20. $2,133.33 21. $386.55 22. $1,131.00 23. 20%
24. 67.1% 25. $9.45 26. $1,380 27. $2.86 28. $936.00
29. $5,572.50 30. $3,010 31. $687.50

CHAPTER 4, Form B 1. $255.00 2. $210.25
3. $352.80 4. $209.59 5. $226.88 6. $171.88 7. $52.14
8. $61.46 9. $191.25 10. $13 11. $437.25 12. 189 h
13. $311.60 14. $7,342.65 15. $96.16 16. $1,231.42
17. $178.37 18. $480.77 19. $17,850 20. $28,400
21. $337.52 22. $1,375.40 23. 1.9% 24. 72.8%
25. $9.45 26. $1,287.00 27. $29.00 28. $213.50
29. $1,912.50 30. $2,679.00 31. $988.00

CHAPTER 5, Form A 1. $57.56 2. $2.44 3. $63.70
4. $6.30 5. $17.00 6. $3.00 7. $0.55 8. $4.05
9. $710.95 10. $121.85 11. $773.72 12. $239.68
13. $0.50 14. $0.48 15. $2.78 16. $5.11 17. 6
instructors 18. 5 packages 19. 4 packages 20. $\frac{5}{6}$

CHAPTER 5, Form B 1. $20.50 2. $4.50 3. $37.10
4. $2.90 5. $25.75 6. $4.25 7. $39.20 8. $10.80
9. $697.50 10. $594.90 11. $158.80 12. $1,406.95
13. $0.78 14. $0.48 15. $2.43 16. $5.74 17. 4 classes
18. 57 packages 19. 3 packages 20. $\frac{2}{3}$

CHAPTER 6, Form A 1. $89.99 2. $186.15 3. 23%
4. $9.00 5. $195.68 6. $0.78 7. $16.43 8. $20.89
9. $266.64 10. 16.1¢/oz 11. 15.6¢/oz 12. Box B 13. no
correct 14. correct 15. not correct 16. $13.90
17. $9.97 18. $14.79 19. $31.27 20. $40.57 21. $79.69

CHAPTER 6, Form B 1. $194.99 2. $295.96 3. 33%
4. $78.75 5. $134.49 6. $0.79 7. $20.54 8. $146.36
9. $2,398.11 10. 18.8¢/oz 11. 18.3¢/oz 12. Meat B
13. correct 14. not correct 15. not correct 16. $12.59
17. $11.41 18. $6.50 19. $14.08 20. $22.94 21. $24.24

CHAPTER 7, Form A 1. 275 2. July 17, 1988
3. Forty-seven and 50/100 4. the signature 5. ⁻24.15
6. 518.48 7. ⁺264.50 8. 782.98 9. ⁻103.60
10. 679.38 11. ⁻72.42 12. 606.96 13. $575.90
14. $459.39 15. $939.96 16. $227.54 17. $409.91
18. $579.65 19. $513.79 20. $1,058.61 21. $112.50
22. $348.75 23. $168.75 24. $1,470.00 25. $1,902.30
26. $3,546.50 27. $8,315.58 28. $8,867.60

CHAPTER 7, Form B 1. 025 2. May 9, 1988 3. 32.15
4. the signature 5. ⁻14.90 6. 89.61 7. ⁻25.00 8. 64.61
9. ⁻52.72 10. 11.89 11. ⁺212.49 12. 224.39
13. $727.74 14. $938.59 15. $380.03 16. $565.75
17. $264.89 18. $285.50 19. $125.12 20. $690.34
21. $78.75 22. $168.75 23. $696.00 24. $1,716.00
25. $5,632.50 26. $9,926.20 27. $16,811.25
28. $19,105.60

CHAPTER 8, Form A 1. $6.41 2. $660.27 3. $2.43
4. $225.87 5. $6.31 6. $648.46 7. $13.22 8. $1,055.38
9. $0.17 10. $2.35 11. $0.41 12. $1.26 13. $140.60
14. $16,872.00 15. $6,556.40 16. $21,356.40

17. $462.00 **18.** $62.00 **19.** $450.94 **20.** $83.98
21. $2.20 **22.** $3.95 **23.** $614.40 **24.** $4,648.44

CHAPTER 8, Form B **1.** $3.10 **2.** $289.65 **3.** $0.68
4. $104.74 **5.** $6.64 **6.** $622.57 **7.** $17.59 **8.** $1,663.27
9. $0.49 **10.** $1.18 **11.** $3.58 **12.** $1.92 **13.** $183.04
14. $10,982.40 **15.** $16,194.40 **16.** $33,044.40
17. $126.00 **18.** $27.02 **19.** $319.64 **20.** $56.84
21. $2.52 **22.** $3.26 **23.** $1,083.75 **24.** $1,866.24

CHAPTER 9, Form A **1.** $11,203.20 **2.** $14,095
3. $8,221.25 **4.** $13,371.80 **5.** 30 mpg **6.** 20 mpg
7. 10 mpg **8.** 15 mpg **9.** $469.76 **10.** $421.71
11. $491.47 **12.** $522.77 **13.** $364.17 **14.** $424.37
15. $301.19 **16.** $105.27 **17.** $1,290 **18.** $904
19. $1,372 **20.** $1,160 **21.** $348 **22.** $204 **23.** $514.20
24. about $100 **25.** about $200 **26.** about $40 **27.** about
$70

CHAPTER 9, Form B **1.** $12,305 **2.** $10,366.50
3. $13,317.50 **4.** $13,814 **5.** 20 mpg **6.** 20 mpg
7. 40 mpg **8.** 30 mpg **9.** $493.13 **10.** $520.57
11. $454.21 **12.** $518.98 **13.** $952.85 **14.** $205.48
15. $306.48 **16.** $247.62 **17.** $1,935 **18.** $2,490
19. $532 **20.** $978 **21.** $234 **22.** $218.70 **23.** $563.50
24. about $60 **25.** about $30 **26.** about $50 **27.** about
$40

CHAPTER 10, Form A **1.** 275 **2.** 171 to 68 **3.** about 2
mi **4.** about 1 mi **5.** about 4 min **6.** 14 h 15 min **7.** 3 h
45 min **8.** 56 min **9.** 35 min **10.** 10 h 55 min **11.** 1:43
P.M. **12.** 4 min **13.** 3 h 18 min **14.** 9 min **15.** $2,500
16. $0.90 **17.** $6.35 **18.** $15.15 **19.** $9.40 **20.** $11.00
21. $600 **22.** $52 **23.** $1,365 **24.** $1,800

CHAPTER 10, Form B **1.** about 5 mi **2.** about 2 mi
3. Holladay **4.** 306 **5.** about 10 min **6.** 3 h 48 min **7.** 3 h
8 min **8.** 35 min **9.** 5 h 35 min **10.** 1 h 10 min **11.** 11:37
A.M. **12.** #122, #634, #110, #214 **13.** Central 244 **14.** 25
min **15.** $2,200 **16.** $1.10 **17.** $13.05 **18.** $2.90
19. $5.35 **20.** $7.25 **21.** about $780 **22.** about $460
23. about $895 **24.** about $1,015

CHAPTER 11, Form A **1.** $20,927.50 **2.** $3,109.14
3. $451.86 **4.** $1,593.72 **5.** $17,911 **6.** $13,699
7. $1,979 **8.** $63,378 **9.** $4,849 **10.** $1,099 **11.** $0
12. $5,431 **13.** $0 **14.** $549 **15.** $520 **16.** $0 **17.** $94
18. $1,162 **19.** $175 **20.** $0 **21.** $221.63 **22.** $10.37
23. $0 **24.** $601 **25.** $0 **26.** $99.50 **27.** $900 **28.** $300
29. $62.50 **30.** $1,100

CHAPTER 11, Form B **1.** $19,875.12 **2.** $3,008.46
3. $402.59 **4.** $1,473.26 **5.** $29,362 **6.** $17,322
7. $2,479 **8.** $94,460.94 **9.** $4,429 **10.** $1,347 **11.** $0
12. $7,489 **13.** $0 **14.** $812 **15.** $1,308.25 **16.** $93.25
17. $0 **18.** $863 **19.** $46 **20.** $0 **21.** $287.63 **22.** $0
23. $93.37 **24.** $987.93 **25.** $79.07 **26.** $0 **27.** $1,000
28. $550 **29.** $237.50 **30.** $600

CHAPTER 12, Form A **1.** $126 **2.** $161 **3.** $357.84
4. $520.24 **5.** $59,120 **6.** $65,050 **7.** $14,280
8. $40,392 **9.** $11,295 **10.** $86,595 **11.** $35,948
12. $117,648 **13.** $823 **14.** $652 **15.** $1,105 **16.** $1,607
17. $56,100 **18.** $2,120.58 **19.** $42,924 **20.** $2,579.73
21. $80.68 **22.** $66.06 **23.** $70.85 **24.** $266.73
25. $47,875 **26.** $25,500 **27.** $2,920 **28.** $3,270 **29.** 3
gal **30.** 2 gal **31.** $4,320 **32.** $3,900 **33.** $746.30
34. $937.40 **35.** $1,300 **36.** $1,625

CHAPTER 12, Form B **1.** $182 **2.** $249.20 **3.** $427.84
4. $651.56 **5.** $43,940 **6.** $59,940 **7.** $74,880
8. $43,920 **9.** $9,540 **10.** $89,040 **11.** $22,560.12
12. $119,760.12 **13.** $717 **14.** $362 **15.** $1,257
16. $885 **17.** $77,000 **18.** $4,150.30 **19.** $27,390
20. $1,506.45 **21.** $63.07 **22.** $51.48 **23.** $79.68
24. $210.05 **25.** $15,630 **26.** $19,500 **27.** $3,700
28. $1,225 **29.** 1 gal **30.** 2 gal **31.** $3,146 **32.** $4,320
33. $767.52 **34.** $933.94 **35.** $1,220 **36.** $3,479

CHAPTER 13, Form A **1.** $90,000 **2.** $67,500
3. $9,600 **4.** $20,000 **5.** $116 **6.** $65 **7.** $816 **8.** $24.30
9. $6\frac{1}{8}$ in. **10.** $20\frac{3}{8}$ in. **11.** $4\frac{1}{8}$ in. **12.** 2 in. **13.** 4 100-ft
coils **14.** 8 100-ft coils **15.** 3 100-ft coils **16.** 26 100-ft
coils **17.** $535.50 **18.** $351.00 **19.** $542.50 **20.** $424.50
21. 234 square ft **22.** 243 square ft **23.** 247 square ft
24. 387 square ft **25.** 26 trees **26.** 18 sections **27.** a
bush crop **28.** 5 times

CHAPTER 13, Form B **1.** $120,000 **2.** $60,000
3. $19,500 **4.** $18,000 **5.** $142.89 **6.** $39.60 **7.** $207.50
8. $36.00 **9.** $2\frac{1}{2}$ in. **10.** $6\frac{5}{8}$ in. **11.** $10\frac{1}{8}$ in. **12.** $6\frac{5}{16}$ in.
13. 4 100-ft coils **14.** 9 100-ft coils **15.** 6 100-ft coils
16. 29 100-ft coils **17.** $468.75 **18.** $321 **19.** $230.13
20. $231.25 **21.** 233 square ft **22.** 438 square ft **23.** 301
square ft **24.** 265 square ft **25.** 14 bushes **26.** 80 ft
27. 6 bushes **28.** mixed

CHAPTER 14, Form A **1.** $196.30 **2.** $420 **3.** $450
4. $2,135 **5.** $521.74 **6.** $32 **7.** 6,642 francs **8.** 19,785
yen **9.** 14 items **10.** 1,750 yo-yos **11.** 33 items **12.** 180
maps **13.** 3 mL **14.** 1.5 mL **15.** 4 tablets **16.** $2\frac{1}{2}$ tablets
17. 6:32 **18.** $2,340 **19.** $993 **20.** $741 **21.** about 160
students **22.** 1985 **23.** 1986 and 1987 **24.** The numbers
decreased.

CHAPTER 14, Form B **1.** $376.13 **2.** $306.72
3. $3,072 **4.** $196 **5.** 458 pounds **6.** 354 Canadian
dollars **7.** $499.27 **8.** $3.75 **9.** 375 items **10.** 400 plates
11. 48 items **12.** 150 books **13.** 2.5 mL **14.** 5 mL **15.** 3
tablets **16.** 2.5 tablets **17.** 5:24 **18.** $400 **19.** $1,140
20. $1,360 **21.** 35 cars **22.** March **23.** April
24. January

CHAPTER 15, Form A **1.** 50% **2.** 10% **3.** 2.4%
4. 11.1% **5.** $4,850 **6.** $16,512 **7.** $29,211 **8.** $27,303
9. $459 **10.** $65 **11.** $25 **12.** $24 **13.** $511 **14.** $1,877
15. $55 **16.** $312 **17.** $1,399 **18.** $83 **19.** $456
20. $830.50 **21.** $25,778 **22.** $11,424 **23.** $14,354
24. $84,725 **25.** $276,830 **26.** $3,830 **27.** $59,000
28. $397,130

CHAPTER 15, Form B **1.** 50% **2.** 25% **3.** 9.1%
4. 4.5% **5.** $218.50 **6.** $498.40 **7.** $13,299 **8.** $27,830
9. $73 **10.** $109 **11.** $43 **12.** $59 **13.** $304 **14.** $1,538
15. $93 **16.** $493 **17.** $2,028 **18.** $22 **19.** $562
20. $412.50 **21.** $202,431 **22.** $84,976 **23.** $117,455
24. $18,191 **25.** $331,390 **26.** $16,330 **27.** $469,450
28. $93,980

CHAPTER 16, Form A **1.** 2.5% **2.** 7.5% **3.** 4.9%
4. 6.5% **5.** $583.20 **6.** $131.68 **7.** $418.60 **8.** $389.50
9. $16.20 **10.** $140.76 **11.** $198.00 **12.** $862.50
13. $4,625 **14.** $3,762.50 **15.** $12,750 **16.** $962.50
17. $2,250 **18.** $750 **19.** $400 **20.** $60.35 **21.** $75.91
22. $713.80 **23.** $482.30 **24.** $5,850 **25.** $7,500
26. $12,000 **27.** $5,375 **28.** $20,000 **29.** $30,000
30. $25,000 **31.** the mutual fund **32.** $8.80 **33.** $8.75
34. the municipal bond

CHAPTER 16, Form B   1. 17.8% 2. 7.7% 3. 10%
4. 9.3% 5. $126.16 6. $525 7. $3,614 8. $150.60
9. $86.88 10. $229.77 11. $386.50 12. $9,750
13. $2,212.50 14. $10,500 15. $3,825 16. $3,250
17. $625 18. $3,400 19. $118.75 20. $69.13
21. $78.95 22. $640.80 23. $729.50 24. $9,650
25. $8,750 26. $6,285 27. $17,136 28. $16,000
29. $40,000 30. $15,000 31. the CD 32. $11.50
33. $50 34. the mutual fund

CHAPTER 17, Form A   1. $\frac{1}{6}$ 2. $\frac{3}{4}$ 3. $\frac{1}{4}$ 4. $\frac{7}{12}$ 5. $\frac{1}{16}$
6. $\frac{1}{9}$ 7. $\frac{5}{24}$ 8. $\frac{1}{2}$ 9. $\frac{1}{11}$ 10. $\frac{1}{22}$ 11. $\frac{9}{44}$ 12. $\frac{5}{33}$ 13. (1,H);
(1,T); (2,H); (2,T); (3,H); (3,T); (4,H); (4,T) 14. (H,H); (T,H);
(H,T); (T,T) 15. (R,X); (R,Y); (G,X); (G,Y); (B,X); (B,Y)
16. a 17. 18 18. 6 19. 24 20. 300 21. 150 spins 22. 45
tails 23. 200 5s 24. 40 Achilles 25. J-C, A-RB; J-C, A-C;
J-RB, A-RB; J-RB, A-C 26. ABC; BCA; CAB; ACB; BAC;
CBA 27. J,M; J,C; M,C; J,A; A,M; A,C 28. D,J; D,R; J,R;
L,R; D,W; J,W; W,L; W,R; D,L; J,L

CHAPTER 17, Form B   1. $\frac{1}{4}$ 2. $\frac{9}{10}$ 3. $\frac{3}{20}$ 4. $\frac{13}{20}$ 5. $\frac{1}{19}$
6. $\frac{1}{38}$ 7. $\frac{9}{38}$ 8. $\frac{3}{38}$ 9. $\frac{7}{40}$ 10. $\frac{51}{400}$ 11. $\frac{1}{100}$ 12. $\frac{1}{8}$
13. C,1; C,2; C,3; T,1; T,2; T,3 14. (R,1); (R,2); (R,3);
(R,4); (R,5); (R,6); (B,1); (B,2); (B,3); (B,4); (B,5); (B,6)
15. (H,R); (T,R); (H,Y); (T,Y); (H,B); (T,B); (H,G); (T,G)
16. (A,4); (A,5); (A,6); (B,4); (B,5); (B,6); (C,4); (C,5); (C,6)
17. 360 18. 60 19. 250 20. 21 21. 40 times 22. 140
times 23. 25 24. 5 25. D,M; D,T; M,T; D,L; M,L; L,T
26. A,L; H,L; A,H 27. K-c, J-c; K-t, J-c; K-c, J-t; K-t, J-t
28. RST; STR; RTS; TRS; SRT; TSR

CHAPTER 18, Form A   1. 21 2. 10 3. 17 4. 3 5. $\frac{12}{4}x$
6. 5(3 + x) 7. 24 = 20 + x 8. 63 = 9(x + 1) 9. 11 10. 10
11. 2 12. 5 13. $\frac{n}{2}$ = 7; 14 14. 96 = 12b; 8 boxes
15. 24 = n − 8; 32 children 16. 4 + c = 10; 6 children
17. 34 = 2m − 10; 22 marbles 18. 3x + 2 = 14; 4 cakes
19. $75 + .2x − $25 = $80; $150 20. 15 = 3 + 4a; 3
years 21. $19,250 22. $1,584 23. $1,733.40
24. $18.72

CHAPTER 18, Form B   1. 15 2. 27 3. 5 4. 2
5. (x + 10) − 3 6. $\frac{x}{2}$ − 7 7. 6(x − 2) = 30
8. 50 = 10x 9. 99 10. 26 11. 9 12. 13 13. 72 = 9r;
8 rows 14. $\frac{n}{4}$ = 16; 64 15. 13 = 3t − 5; 6 people
16. 143 = 20 + t; 123 cards 17. 84 = 34 + 5s; 10 coins
18. 32 = 2j + 4; 14 years 19. 29 = 5 + 2s; 12 stocks
20. 144 = 12 + $\frac{m}{2}$; 264 cookies 21. $112,910
22. $3,712.50 23. $3,544 24. $535

Final Test

Part A   1. a 2. a 3. d 4. b 5. c 6. b 7. d 8. a 9. c
10. b 11. b 12. d 13. a 14. c 15. c 16. a 17. d 18. a
19. d 20. a 21. b 22. a 23. b 24. d 25. a **Part B**
26. b 27. a 28. b 29. c 30. b 31. a 32. b 33. a 34. d
35. a 36. b 37. c 38. c 39. b 40. c **Part C** 41. b
42. a 43. c 44. d 45. a 46. b 47. b 48. b 49. d 50. a
51. a 52. a 53. c 54. a 55. b 56. d 57. a 58. c 59. b
60. a 61. d 62. c 63. a 64. b 65. a 66. c 67. a 68. d
69. a 70. a 71. b 72. c 73. c 74. b 75. b 76. c 77. d
78. a 79. d 80. c